I0032789

PHOSPHORESCENCE

OR, THE

EMISSION OF LIGHT

BY

MINERALS, PLANTS, AND ANIMALS.

BY

T. L. PHIPSON, Ph.D., F.C.S.,

MEMBER OF THE CHEMICAL SOCIETY OF PARIS, ETC.

LONDON:

LOVELL REEVE & CO., HENRIETTA STREET, COVENT GARDEN.

1862.

[The Right of Translation is reserved.]

Dedication.

———◆———

TO SIR WILLIAM SNOW HARRIS, F.R.S.,
ETC. ETC.

My dear Sir,

I have great pleasure in dedicating this Work to you, as a *souvenir* of the pleasant hours passed in your society when you last visited Paris. If it meet with a fraction of the success which has attended your admirable contributions to Electrical Science, I shall feel justified in having inscribed it to you.

Believe me,

My dear Sir,

Yours very sincerely,

T. L. PHIPSON.

PREFACE.

THE original sketch of this work appeared for the first time in the 'Journal de Médecine et de Pharmacologie,' of Brussels, at the commencement of the year 1858. It was soon after reprinted and published anew at Brussels and Paris. At the same time, my original paper was reproduced in Belgium, without my knowledge, in the 'Revue Populaire des Sciences,' by M. Husson, and in France, in the 'Ami des Sciences,' a weekly paper, edited by M. Victor Meunier.

Unknown to me, it was translated also into German, by Dr. Müller, of Berlin, about two months after its first appearance; and I have since learnt that an Italian edition was expected to appear shortly.

I was thus convinced that the subject of Phosphorescence deserved to be treated far more extensively than had been done in my *brochure*,—a mere sketch, which is now out of print; so that, instead of republishing it with additions, I have completely remodelled the work, and brought forward in the present volume every case of Phosphorescence which it has been in my power to obtain (many of which have originated in my own laboratory), after seeking for and studying the phenomenon in the whole domain of Nature. My attention was first called to this extremely interesting class of natural facts by my physical and chemical studies. They have occupied my thoughts for some time past; and I was the more anxious of treating this subject *in extenso*, since, up to the present day, it has been impossible to give a satisfactory explanation of phosphoric phenomena.

Phosphorescence, indeed, whether manifested by the glowworm, the Bologna stone, a fungus, or a falling star, is generally looked upon as an unexplained and mysterious production of light. I hope, nevertheless, that I have been able to extricate it a little from the obscurity in which it has hitherto been enveloped.

In order to appreciate every circumstance con-

nected with phosphoric phenomena, or the spontaneous emission of light by natural or artificial
substances, as well as by living and dead organic
bodies, these phenomena must be studied in the
whole domain of Nature—in the mineral, in the
vegetable, and in the animal world. I have therefore brought forward what I know on the subject,
with regard to mineral substances, and have extended my investigations to vegetables, to animals,
and to organic matter deprived of life. The phenomenon of Phosphorescence will thus have been
studied simultaneously throughout Nature, and
this is, to my knowledge, the first time that the
interesting series of facts I have consigned to
these pages have been looked upon as constitu-
ting a whole.

Phosphorescent properties, although developed
to a prodigious degree in the insect world, are
found nevertheless to exist in numerous other animals, in many plants, and also in certain minerals
and chemical products. They pertain at once to
the science of chemistry and physics, as well as to
botany and physiology. Those who possess a
profound knowledge of these different branches of
natural history, can alone hope to arrive at the

cause of the varied phosphoric phenomena with which observation has already furnished us, or to explain these phenomena in a satisfactory manner. A flame is always a flame, light is everywhere light; but it remains necessary to ascertain *how this light is produced* in the different circumstances under which it is observed. I myself do not pretend to have snatched from Nature the entire secret of Phosphorescence, but I have reasons to hope that the observations contained in this work will prove, that, owing to the rapid progress that natural science has made during the present century, I have been able to tread in a firmer path than many who have preceded me, and that I have penetrated a little way into the track which will conduct us finally to the desired goal.

My work is essentially divided into four parts. The first treats concisely of mineral phosphorescence; it includes also the history of certain meteorological manifestations of light, some of which are extremely remarkable. In the second, I have said what I know of the emission of light by plants and vegetable substances; and have proceeded in the third division to investigate the phenomenon of Phosphorescence in dead animal substances, and the emission of light by living animals.

The fourth division comprises some historical notes upon the subject, and the theory by which I have endeavoured to account for the various phenomena to which the attention of the reader is called in the previous chapters. I have purposely placed my theoretical considerations in this section, as they will thus have a better chance of fair appreciation than if their elements had been dispersed throughout the work, appended to every isolated fact or experiment as mentioned or described.

CONTENTS.

Part Second.

PHOSPHORESCENCE OF VEGETABLES.

Part Third.

PHOSPHORESCENCE OF ANIMALS.

Part Fourth.

HISTORICAL, THEORETICAL, AND PRACTICAL CONSIDERATIONS.

APPENDIX.

THE PHOSPHORESCENCE

OF NATURAL OBJECTS.

INTRODUCTION.

ABOUT the latter end of the sixteenth century there lived in a narrow, winding street of the old town of Bologna, a certain cobbler, Vincenzo Cascariolo,* who devoted much of his time to alchemy. Some say that he even quitted his trade, and applied himself exclusively to chemical labours, but I am inclined to doubt the fact. However cheap living might then have been in Italy, alchemy would indeed have formed a bad substitute for the last in many respects. In spite of this, Signor Vincenzo was so bent upon making gold, that his little workshop contained nearly all the mysterious chemical apparatus of the day. Phials, retorts, and crucibles found room among awls, lasts, and leather; and Vincenzo Cascariolo,

---

* Some write his name Casciarolo.

no doubt, looked upon shoe-mending as a sorry occupation for a man initiated in the secrets of the " sublime art," and who might indeed have ranked among the first adepts of his day.

Some time had elapsed since our cobbler had set his heart upon gold-making, when, strolling one fine Sunday evening near the little eminence known as the Monte Paterno, about a league from Bologna, he picked up a stone, similar to any other stone, save perhaps in one particular, its great weight.

The fact struck him. This stone possesses, thought he, one of the properties of gold. Perhaps he imagined that it contained gold which might be extracted; or, may be, he fancied it would be capable, from its heaviness, of transforming vile or imperfect metals into gold, by imparting to them its characteristic property.

Cardan, Van Helmont, Libavius, and many other distinguished alchemists, had lived before Cascariolo's time, but I know not whether he studied their works, and I doubt whether he would have profited much by them if he had.

It is impossible to ascertain therefore what prominent idea, or what kind of theory reigned in the cobbler's mind on the discovery of this stone, destined to become celebrated and to immortalize his name. However, no sooner had he collected a certain number of specimens, than he hastened back to his little workshop, and began immediately to experimentize upon the mineral.

It appears most probable that Cascariolo looked upon the sulphate of baryta, or heavy-spar,—for such was the object of his curiosity,—as a metallic ore, and supposed that by heating it with charcoal in a hot fire, he would be able to extract a metal —perhaps gold! His hopes in this respect were not realized, but he nevertheless succeeded in obtaining one of the most curious of substances,— a body which, to use the words of an old physicist, "absorbs the rays of the sun by day, to emit them by night."

At this period there was at Bologna a well-known alchemist, Scipio Begatello, who had rendered himself remarkable by his attachment to the art of gold-making; and in the year 1602 the cobbler brought to him the product of his experiments, showed him the substance produced by calcination (and which he called by the mystical name of *lapis solaris*), and endeavoured to convince Begatello that from the weight of the stone which had furnished it, from its power of attracting and retaining the golden light of the sun, this shining substance would doubtless be fit for converting the more ignoble metals into gold—the *sol* of the alchemists. He showed it also to Maginus, a distinguished professor of mathematics, who, being no adept, did not keep the matter a secret, as Begatello seems to have done, but sent both the mineral and the substance produced from it, to the princes and learned men of the day, thereby contributing more than any other person to make known this singular discovery.

The stone discovered by Cascariolo is now known as Barytine, or Heavy-spar (sulphate of baryta). By heating it with charcoal he had transformed it into sulphuret of barium, a substance which has the curious property of shining in the dark, after it has been exposed for some time to the rays of the sun.

Many years after its discovery, the German chemist Marggraf found an easy and certain method of preparing it, by making into a paste with water a mixture of pulverized barytine and flour, and submitting the whole to heat in a closed crucible. The sulphuret thus produced is placed in a well-corked glass jar, or made into stars, which shine marvellously in the dark after they have been exposed to the sun for a short time.

Such is the history of the discovery of the substance first known to be phosphorescent by insolation. For many years it has been sold in the streets of Bologna as a curiosity, under the name of Solar Phosphorus, or the Bologna Stone. Marggraf showed that other minerals, other varieties of heavy-spar, were capable of furnishing similar "light magnets" or "luminous stones," and at the present day a great number of substances are known to possess the same singular property.

In 1663, the celebrated English chemist Robert Boyle announced to the world that the Diamond possessed the same luminous property as the Bologna Stone; and what we have just told of Cascariolo's labours has its parallel in the discovery of another luminous body, far more remarkable than either.

In the seventeenth century there lived at Hamburg an alchemist named Brandt, who having endeavoured for many years, but in vain, to convert other metals into gold, was struck one day with the golden colour of urine, and doubted not but that this liquid contained some substance that would realize his dreams. Brandt thought that by acting upon the metals he wished to convert into gold with a blackish extract he had prepared by concentrating and evaporating urine, he would certainly operate the desired transmutation. He therefore introduced this black extract into a retort with the metals in question, lighted his furnace, and watched intently the progress of the operation. The result was negative: the metals after the experiment remained unchanged. However, one evening, after having distilled a considerable quantity of urine over some metal or other, and having pushed the distillation as far as possible, he was surprised to find that a peculiar shining body had passed over into the recipient. He repeated the experiment with the black extract, and, after applying as great a heat as he could muster, he obtained a notable quantity of this trange shining substance, collected it in silent astonishment, studied its properties, found that it was extremely inflammable, and that it possessed the curious property of shining intensely in the dark.

These experiments were made in the year 1669, and the luminous substance was called Phosphorus.

Brandt immediately sent a specimen of this wonderful body to Kunkel, chemist to John George II., Elector of Saxony, and one of the most distinguished *savants* of the day, but did not disclose to him the secret of its preparation. Kunkel, in his turn, showed it to his friend Kraft, of Dresden, who found it so marvellous that he proposed to set out immediately for Hamburg, and endeavour to discover how this luminous substance was prepared. He took two hundred dollars with him, and succeeded in buying for that sum the whole detail of the preparation. But Brandt only delivered it on the condition that Kraft should disclose it to no one.

Kunkel, whose passion for chemistry was intense, felt such disappointment when he learnt that Kraft possessed the secret, and yet could not make it known to him, that he determined to set about finding it out for himself. He knew that Brandt, who died shortly afterwards, had devoted most of his life to experiments on urine, and he felt convinced that phosphorus must have been obtained from that liquid. After many and varied experiments, quite unsuccessful, Kunkel at last obtained (in the year 1674) the substance he sought after so long and so obstinately.*

In the year 1675, another chemist, Baudoin, prepared a new "phosphorus," or shining substance, by calcining nitrate of lime. Since then, many substances which shine in the dark after exposure to the sun, have been discovered. One of the most remarkable, perhaps, is "Canton's Phosphorus," or sulphuret of calcium, obtained according to the author just named " by heating a mixture of three parts of sifted calcined oyster-shells with one part of sulphur to an intense heat for one hour." It can also be prepared by calcining plaster of Paris with common charcoal.

The peculiar and sometimes extremely vivid phosphorescence of the sea was known in antiquity. Pliny speaks of it, and of the phosphorescence of certain Medusæ. But it was not till long afterwards that the cause of this wonderful phe-

* Some authors state that phosphorus was discovered in England, about the same time, by Robert Boyle. In 1769, Gahn, the celebrated Swedish mineralogist, discovered it in bones, and published, with the illustrious Scheele, a new process for extracting it, which is similar to the one practised at the present day.

nomenon was ascertained. The names of the authors of this discovery, and the dates of their investigations, are given in this work, in their proper places. The same observation will apply to those singular cases of phosphorescence which have been observed in the vegetable kingdom, upon flesh, decayed wood, etc.

It will be easily understood what is meant by the term *Phosphorescence*, when we remind our readers that phosphorus, which shines so curiously in the dark, and which enters into the composition of our common lucifer matches, is the most remarkable of all phosphorescent bodies. The word "phosphorus," which signifies a substance that bears or emits a light, has frequently been applied to various other substances besides the non-metallic element termed *phosphorus* in chemistry, on account of the property these substances possess likewise of shining in the dark.

PART I.

MINERAL PHOSPHORESCENCE.

CHAPTER I.

PHOSPHORESCENCE AFTER INSOLATION.

SEVERAL substances manifest the strange property of emitting light when they are placed in darkness, after having been exposed for some time to the direct rays of the sun. In some cases a very short exposure to sunlight is sufficient to excite the manifestation of this remarkable property, and in others the direct rays of the sun are not necessary : it suffices that the substance experimented upon be exposed to the dull light of a cloudy day. To this phenomenon the denomination of *Phosphorescence after insolation* has been given.

The substances which possess this property in the highest degree are the Bologna stone, or solar phosphorus, certain varieties of fluor-spar and carbonate of lime, some fossils, calcined shells or pearls, phosphate of lime, arseniate of lime, etc. Many diamonds shine with brilliancy in the dark if they have previously undergone an exposure of some seconds' duration only to solar light. But

no substance surpasses in this respect sulphuret
of barium.

It is now a long time since the cobbler of Bologna,
in Italy, astonished and amused his friends with a
peculiar substance since known as *Bologna phosphorus, Bologna stone,* or *Solar phosphorus,* which
shines brightly in the dark after having been placed
in the sunlight for some time. This substance is
sulphuret of barium. The cobbler prepared it by
heating red-hot with charcoal a piece of *sulphate
of baryta,* or *Barytine,* (Fig. 1,) a stone which he

Fig. 1.

picked up in the secondary strata of the Monte
Paterno, where he found it in lumps of considerable
weight.* The German chemist, Marggraf, used to
prepare solar phosphorus by powdering down the
stone, and making it into thin cakes, with a mixture of flour and water, before submitting it to
calcination. This "Bologna phosphorus" was the
first substance known to become phosphorescent
after insolation, and, consequently, it has been

* Barytine is found abundantly in Derbyshire, Cumberland,
the Isle of Arran, etc.

submitted to many and varied experiments. It is best obtained by the calcination of pulverized sulphate of baryta, made into a firm paste with common gum. It should be preserved in a bottle which closes hermetically with a glass stopper.

When such a bottle and its contents are exposed to the rays of the sun, or even to daylight, for a certain time, and then taken into a dark room, the sulphuret of barium is seen to be beautifully phosphorescent, or to shine like common phosphorus, and the phenomenon will sometimes last a whole hour. The most intense cold does not affect this phosphorescence, and it manifests itself precisely in the same manner whether the sulphuret be placed *in vacuo* or in the air.

When nitrate of lime is melted for ten minutes in a crucible, it leaves a residue which manifests, to a less extent, the same phosphorescent property. This residue, which is nothing more than pure lime, or a mixture of lime and nitrite of lime, was known for some time as *"Baudouin's phosphorus."* A like phenomenon is observed with calcined shells.

Sulphuret of calcium possesses the same phosphorescent qualities as the sulphuret of barium alluded to above; hence the former is sometimes known as *"Canton's phosphorus."* Canton prepared it by heating a mixture of three parts of sifted calcined oyster-shells, with one part of sulphur, to an intense heat for one hour.* It can also be formed by heating gypsum with charcoal.

Some diamonds, but not all, possess the same property, and many other substances have been observed to be more or less phosphorescent in the same circumstances, that is, after insolation.†

Landrin (*Dict. de Min.*) asserts that *radiated sulphate of baryta*, certain *natural fluorides, rocksalt, amber (succinum)*, and *quartz*, become luminous for a few instants after they have been exposed to the sun.

Walls that have been painted or whitewashed with *lime*, are apt to become luminous at night after they have received the action of the sun's rays in the daytime. Whitewashed houses are, on account of their phosphorescent quality, visible at a great distance on the darkest nights.

It was natural enough that the action which the coloured rays of the solar spectrum exercise upon these substances, that become phosphorescent *after insolation*, should be early investigated; and in 1775, Wilson published his ' Series of Experiments on the Phosphori,' in which paper he asserts that the most refrangible rays of the solar spectrum determine a vivid phosphorescence in sulphuret of calcium (" Canton's phosphorus"), whilst those rays which are the least refrangible—*i.e.* those situated near the *red* light of the spectrum—cause

* Philosophical Transactions, 1768.   † See Chapter VI.

the phosphorescence excited by the other rays to cease! Ritter was also aware of this, and about the same time Beccaria found that " the violet rays of the spectrum are the most apt, the red rays the least apt, to develope phosphorescence in solar phosphori." Becquerel affirms also, from his own experiments, that the property possessed by light of rendering certain bodies luminous in the dark, appears to reside—if not entirely—at least to a great extent in the violet rays, whilst the red rays are completely deprived of this property, a fact noted also by Heinrich.

Biot, Arago, Daguerre, and others, have made many researches on this subject. They have shown, among other curious facts, that with the invisible rays—sometimes termed chemical or actinic rays—situated underneath the luminous part of the solar spectrum, it is possible to render a phosphorescent body luminous, or at least visible; whilst, when plunged in the visible rays, red, yellow, orange, green, etc., not only this same body is not lighted up, but its light previously excited by the invisible rays is extinguished.

This curious phenomenon has been admirably investigated in England by Professor Stokes, who has denominated it *Fluorescence*, and who has shown that a considerable number of substances, besides those known as *solar phosphori*, act upon these invisible rays of the spectrum and render them visible.

Dessaignes and the elder Becquerel have re-marked that *those bodies which are good conductors of electricity are not phosphorescent after insola-tion*. We shall have occasion to refer again to this important fact. Biot and Becquerel have both proved that electricity acts upon the solar phos-phori in the same manner as insolation. An elec-tric discharge renders them luminous in the dark for some time after the discharge has passed (a discovery originally made by Grothuss, and with which both Canton and Dessaignes were familiar), and they have also shown that differently-coloured rays of light modify this action in the same man-ner as the differently-coloured rays of the solar spectrum. Moreover, phosphori that have lost their phosphorescence, or that have ceased to shine in the dark after a first insolation, recover their luminous property when acted upon by the electric light—an observation we owe to Grothuss and Becquerel—and when this light is passed through certain transparent screens, such as plates of glass, of quartz, or different salts, it is observed that these screens become obstacles and sometimes com-pletely prevent any phosphorescent radiations.

Electric discharges proceeding from a battery communicate a recognizable phosphorescence of a longer or shorter duration, to a great number of bodies *which are bad conductors or non-conductors of electricity*. This phenomenon may be observed, for instance, with sugar, dry chalk, and many other substances.

Among bodies slightly phosphorescent after insolation, we may name melted potash and soda, dry nitrate of lime, and chloride of calcium, sulphate of potash, sulphate of soda, arragonite, calcspar, dolomite, carbonate of strontia, carbonate of baryta, different calcareous earths, phosphates of lime, sulphate of baryta, sulphate of strontia, etc.

According to Ed. Becquerel, other substances are phosphorescent after insolation, but in order to observe it we must remain some time in a dark room, and then, by means of a hole in the shutter, expose the body experimented upon to the light, at the same time keeping the eyes closed until the hole in the shutter is shut again. By experimenting in this manner, many substances are seen to be phosphorescent for a few seconds after insolation; amongst others numerous minerals, salts, organic substances such as paper, gum, sugar, teeth, etc.

Long before Becquerel, however, we find in the article 'Phosphorus," of the 'Encyclopædia Perthensis,' it has been found "that *almost all terrestrial bodies,* upon being exposed to light, will appear luminous for a little while in the dark, metals only excepted."

CHAPTER II.

PHOSPHORESCENCE BY HEAT.

MANY substances become phosphorescent when
they are heated to a certain temperature. Such,
for instance, are fluor-spar, lime, sulphuret of calcium, diamonds, etc. They manifest their phosphoric qualities when, after being pulverized or
broken into small fragments, they are thrown upon
a heated surface.

Fig. 2.

Some varieties of apatite (fig. 2—phosphate of
lime with fluoride or chloride of calcium) are
phosphorescent when heated, especially the
coarser varieties. I have proved that their phosphorescence is owing to the *fluoride of calcium*
which forms part of the mineral, for *pure phosphate of lime* does not show any phosphoric light

when heated; nor do most of the so-called chlorapatites, which contain chloride of calcium substituted in part or wholly for the fluoride.

Of all these substances, the most remarkable is fluor-spar (fluoride of calcium—fig. 3). When

Fig. 3.

thrown in the dark upon heated mercury, into boiling water, or on to a hot shovel, this mineral immediately emits a brilliant phosphoric light. Some specimens possess this property to a greater extent than others. A certain green variety of fluor-spar called *Chlorophane* becomes phosphorescent at the low temperature of 20° or 25° (centigrade), which is almost that of our summers. Rare descriptions of chlorophane become phosphorescent in a dark room from the mere warmth of the hand. According to Landrin (Dict. de Minéralogie) some varieties are almost constantly luminous in the dark.*

* Fluoride of calcium loses its phosphoric property after it has been once heated. Miller asserts ('Elem. of Chemistry') that when a phosphorescent fluoride of calcium is dissolved in hydrochloric acid, and then precipitated by ammonia, the precipitate is

Common salt (chloride of sodium), chloride of mercury, arsenious acid, etc., are phosphorescent only at a temperature of about 200° (centigrade).

White flocconous oxide of zinc may be heated to a very high temperature without melting or volatilizing; but whilst heated it is observed to turn yellow, becoming white again on cooling. Now, whilst this transformation of colour from yellow to white is going on, the oxide of zinc is seen to glow with a faint blue phosphoric light. This change of colour and this emission of light, observes Baudrimont (in his 'Traité de Chimie,' vol. ii.), seem to indicate that the oxide of zinc undergoes what is termed an *isomeric modification* (change of chemical properties) at a high temperature, and returns again to its primitive state on cooling.

Bendant affirms that a crystal of fluor-spar which is very perfect and transparent, will not become phosphorescent by heat until one of its surfaces has been roughened a little on a piece of sandstone; he states also that diamonds which have not been cut are not phosphorescent by heat, but that they become so as soon as they are cut or polished.

---

phosphorescent. This is not the case, however, if the fluoride has been previously heated enough to destroy its phosphorescence. Solution and precipitation have therefore no power to destroy or to restore this curious property.

Almost any substance, whether organic or mineral, if a non-conductor of electricity, becomes more or less phosphorescent when it is thrown upon a molten bath of the easily-melting alloy of D'Arcet. Indeed, Wedgwood published an elaborate paper upon this subject in the 'Philosophical Transactions' for 1792. He experimented upon an extensive variety of substances both mineral and organic, by reducing the body to a moderately fine powder, and sprinkling it by small portions at a time on a thick plate of iron heated just below visible redness, and removing the whole to a perfectly dark place. He has given a prodigious list of substances which appear luminous for a few instants when submitted to this treatment.

When a body which is known to be phosphorescent by heat loses, from some undetermined cause, its phosphoric property, the latter can be restored to it by means of electricity, as we have seen in the foregoing chapter regarding substances which are phosphorescent after insolation. For example, certain diamonds which cannot be made to give out any phosphoric radiation by heat, will do so after one or two electric discharges have been passed over them. This curious fact was made known by the German *savant* Grothuss.

Pearsall (in 'Journal of Royal Institution,' vol. i.) has described experiments proving that a dozen electric discharges passed through non-phosphorescent bodies such as marble, certain varieties of apatite, etc., will give them the property of becoming luminous by heat. On twenty-one days of *exposure to light*, these substances rendered artificially phosphorescent lost that property again. But if kept in a *dark* room, they retained it. The term "non-phosphorescent bodies," used by Pearsall, is rather exclusive; as phenomena of phosphorescence are so universally spread, that scarcely any substance, if properly experimented with, will prove to be non-phosphorescent in the strict sense of the word. His observation is, however, exceedingly remarkable.

Some years ago M. Schönbein showed that metallic *arsenic* becomes phosphorescent when its temperature is raised to a certain degree.* I imagined that *antimony* might present the same phenomenon, but found it was not the case. *Stibine*, or native sulphuret of antimony, I found, however, to be very phosphorescent when heated in a crucible to a dull-red heat. The light produced is white, with a slight tinge of yellow. I have lately observed that *copper*, native *sulphuret of copper*, *gold* and *silver* are notably phosphorescent when melted on charcoal before the blow-

---

* The metallic arsenic is placed in a small glass globe, and heated with a spirit-lamp. The light is emitted at the same time that the characteristic garlic odour is developed.

pipe. As soon as copper is thoroughly melted (at the inner flame), it glows with a greenish-yellow light, similar to that of the glow-worm. On cooling a little, it rapidly loses this property, and at the same time a molecular change is observed on the surface of the metal. I have also found that the mineral *lepidolite* is as brilliantly phosphorescent by heat as fluor-spar. But to observe this phenomenon properly, it should be viewed through a piece of glass coloured blue by oxide of cobalt. In these circumstances, the phosphoric light of lepidolite before the blowpipe is very fine. The blue glass extinguishes the orange-red light of the heated charcoal.

Among organic salts it has been observed that *sulphate of quinine* and *sulphate of chinconine* become phosphorescent under the influence of heat. M. Böttger has remarked that these salts do not shine in the dark as long as their temperature continues to rise; they become phosphorescent only when, after being heated, the temperature begins to decrease and they remain in a luminous state for some minutes whilst cooling. Pure *quinine* and sulphate of quinine are very phosphorescent in this manner: the phosphoric light given out by sulphate of quinine whilst cooling is sufficiently strong to enable one to read by it. Pure *cinchonine* does not appear to be phosphorescent by heat, but sulphate of cinchonine is so, though to a less degree than sulphate of quinine.

Paper is also capable of becoming luminous when heated : fix upon a plate of copper any characters of the same metal; these characters should be about two-tenths of an inch thick. At the back of the plate fix an iron rod terminating in a wooden handle. When the plate is heated, and then violently pressed upon very dry paper lying upon three or four folds of cloth, the characters thus impressed will appear faintly luminous in the dark until the paper is cool.

CHAPTER III.

PHOSPHORESCENCE BY CLEAVAGE, FRICTION, PER-
CUSSION, CRYSTALLIZATION, AND MOLECULAR
OR CHEMICAL CHANGE.

IT has been observed that numerous minerals and
chemical products emit light when they are split,
i. e. during the process of cleavage; others, by
friction; others again, by percussion or whilst
they crystallize, etc.

When a lamina of mica, for instance, is divided
by cleavage, and the operation proceeds in a dark
room, a feeble emission of light is perceived at the
moment the separation of the two plates occurs.
Each of the two plates thus separated is found
afterwards to be electric: the one shows positive
electricity, the other negative electricity. Something similar is observed in the cleavage of feldspar, which, according to Landrin and some other
authors, emits a feeble light in this circumstance,
the phosphorescence lasting only a few moments.

Another instance is afforded by boracic acid.
When boracic acid is melted in a crucible and then

allowed to cool, it cracks or splits up as its temperature decreases, and, at the same time, emits a feeble light. When vanadic acid is melted, it crystallizes on cooling, and during the whole time that this crystallization lasts, the substance glows with a red phosphoric light. When phosphate of lead is melted before the blowpipe, it forms a crystalline bead on cooling, and whilst this crystallization takes place light is produced. I have frequently observed the emission of light by boracic acid: when melted before the blowpipe, and allowed to cool in the dark, it cracks and gives a sudden flash of light when it has cooled for about twenty seconds. The acid should be melted upon a platinum wire bent at the end. Berzelius thought that this light was produced in a similar manner to the electric radiation which is sometimes observed when a card is suddenly torn asunder after being split at one of its corners. Light is also produced when crystals of sugar and nitrate of uranium* are broken. And I have observed the same to take place with *lactine,* or sugar of milk. Another kind of sugar, called *mannite,* presents similar phenomena. An emission of light is also observed when crystals of protochloride of mercury (sublimated calomel) are broken between the fingers.

* The effect is very striking if crystals of nitrate of uranium be shaken up in a bottle.

When chloride of calcium that has been melted in a crucible, is rubbed upon the sleeve in a dark room, it glows with a greenish light. This was first observed by Homberg, hence the name of *Homberg's phosphorus*, by which this substance was formerly known. It is very phosphorescent by percussion.

Certain varieties of blend (sulphuret of zinc) become phosphorescent by percussion and sometimes after very slight friction. Speaking of blend, Dana says :—" Merely the rapid motion of a feather across some specimens of sulphuret of zinc, will often elicit light more or less intense from this mineral."* Other substances require a stronger rubbing, for instance, quartz, flint, etc. In the case of quartz, an odour of ozone is perceived, a fact to which I called attention in the 'Comptes Rendus' of the Paris Academy, in 1860.†　Borax and sugar become luminous also in the dark, when rubbed. Otto de Guericke observed that the globe of sulphur with which he constructed the first electric machine, became luminous when he rubbed it in the dark.

Hawksbee and Picard both discovered that the

* Dana's 'Mineralogy.'

† I find that it requires upwards of two hundred flashes to produce a quantity of ozone equivalent in its effects to one drop of nitric acid. Also, that with white quartz the light is white, but with red quartz or calcined yellow quartz, it acquires a crimson tint, owing to the oxide of iron in the stones.

friction of mercury in the vacuum of the barometer tube produces a phosphorescent light.

When oxygen gas or air are compressed suddenly in a piston, heat alone is produced. The light seen in this experiment is owing to the combustion of some of the oil of the piston, as proved by Thenard. When the piston is imbibed with water, no light is perceived; and M. Saissy, of Lyons, has proved that oxygen alone, by its comburent power, is the cause of this light.

If chlorate of potash, fluor-spar, feldspar, sugar, etc. be struck in the dark, or ground down in a mortar, they present very vivid phosphoric radiations. With crystallized substances which are cleavable, *i. e.* easily divided into thin laminæ, this phosphorescence is very remarkable; with sugar, for instance, each fissure produced by the shock of the pestle gives birth to a streak of light which lasts for an instant, and when a certain quantity of any of these substances is ground down rapidly in a mortar, the whole mass appears as if on fire. This beautiful phenomenon is exceedingly striking when transparent feldspar is experimented upon. It has lately been discovered that dry hypophosphites of lime, soda, etc., become phosphorescent when shaken or stirred in the dark. (*Note.—* These salts are apt to explode violently when evaporated to dryness at too high a temperature. Tuson, in 'Chemical News,' August, 1860.)

In 1858, M. Landerer, of Athens, discovered that an organic salt, *Valerate of quinine,* becomes phosphorescent whilst it is being powdered in a mortar. The light emitted, which is very strong at first, becomes feeble as the pulverization proceeds, and ceases altogether when the crystals are reduced to powder. When the crystallization of fluoride of sodium takes place in a dark room, this salt is seen to twinkle with phosphorescent light. The same is observed when sulphate of soda and sulphate of potash crystallize together. Waechter has observed that chlorate of baryta crystallizes from its solution in long rhombic prisms with production of light.

A most interesting production of light was observed and published ('Journ. des Sc. Physiques et Chimiques,' de M. de Fontenelle), by Professor Pontus, in 1833, who showed that a vivid spark is produced when water is made to freeze rapidly. A small glass globe, terminating in a short tube, is filled with water, the whole is covered with a sponge or cotton-wool imbibed with ether, and placed in an air-pump. As soon as the experimenter begins to produce a vacuum, the ether evaporates, and the sponge or cotton-wool dries, the temperature of the water descends rapidly. But some instants before congelation takes place, *a brilliant spark, perfectly visible in the daytime,* is suddenly shot out of the little tube that terminates the glass globe. M. Pontus has repeated the experiment often, and says that the production of this spark is a sure sign that congelation is about to happen.

It is well known to chemists that arsenious acid exists under two distinct molecular modifications discovered by M. Guibourt, viz. the transparent acid and the opaque acid. Professor H. Rose, of Berlin, has shown that when the transparent variety is dissolved in a hot solution of diluted hydrochloric acid, and the dissolution allowed to cool, the opaque variety is deposited in crystals, and each crystal, as it forms, is accompanied by an emission of light.*

The same emission of light is observed when certain oxides, whilst heated in a crucible to a given temperature, undergo a peculiar molecular change which occasions a modification of their chemical properties. A phosphoric radiation, a sort of incandescence, is remarked the instant that this change takes place. The substances that are remarkably phosphorescent during this molecular change occasioned by heat are, alumina, chromic oxide, oxide of zirconium, tantalic acid, titanic acid, the acids of the new metal niobium, peroxide of iron, and some others. In the mineral gadolinite, this phenomenon is very well

* See Rose's paper on this in the 'Annalen der Physik und Chemie,' 1835.

observed. The rare mineral called samarskite (niobate of iron, of uranium, and of yttria) presents the same phenomenon to a less degree.

Oxide of molybdenum (MoO) is obtained by heating the hydrate of this oxide *in vacuo* : when the water is all evaporated, the dry oxide remaining appears as if on fire; the anhydrous or dry oxide is then completely formed.

According to Berzelius antimoniate of copper becomes luminous when heated, and changes its chemical nature. Something similar has been observed with phosphate of magnesia.

About fifteen years ago, M. Scheerer showed that the density of gadolinite (which, like samarskite, contains no water) *increases* during the phosphoric radiation; the density of the mineral is greater after the experiment than before. M. H. Rose has confirmed this fact anew, and has shown at the same time, that with samarskite the contrary is observed : the density *diminishes*. These observations alone prove that a molecular change takes place during the emission of light.

H. Rose has also confirmed the fact announced some time ago by Regnault as an hypothesis, viz. that the luminous phenomenon which occurs when the substances above alluded to suddenly change their state, is accompanied by an immediate change in their specific heat. Thus, the specific heat of gadolinite diminishes one-fourteenth during the emission of light. This necessitates a disengagement of heat, which is really observed to take place, as is shown in the following experiment:—When chromic oxide, titanic acid, or better than all, gadolinite is heated in a small retort, the end of which, terminating in a capillary tube, plunges into water, the dilated air confined in the apparatus is expulsed through the water uniformly as the heat increases, and when the phenomenon of incandescence takes place, the bubbles of air are, for a moment, driven out violently, indicating a sudden production of heat.

With samarskite and arsenious acid (see p. 19), whose densities *diminish* during the experiment, no heat is disengaged, as with gadolinite. Thus, thinks Rose, when this phosphorescence occurs without any disengagement of heat, it seems to indicate a diminution of density, whilst phosphorescence *with* emission of heat, appears indicative of an *increase* of specific gravity; and "probably, during the *diminution* of density, the caloric is employed to separate the atoms, instead of being disengaged."

These results, obtained at the beginning of the year 1857, are exceedingly remarkable.

In this chapter we should include also the well-known phosphorescence of phosphorus, formerly studied in our ' Recherches nouvelles sur le Phosphore.' The phenomenon occurs when phosphorus

combines with the oxygen of the atmosphere to
form phosphorous and phosphoric acids. As a
chemical phenomenon it is in every respect similar
to the flame produced when potassium or sodium
burns in contact with water, or when many bodies
having very strong affinities for each other, combine, with a production of light. The phosphorescence of phosphorus, and the combustion (formation of phosphorous and phosphoric acids) by
which it is accompanied, occur in air or oxygen
gas at a given temperature. But if the pressure
of these gases be diminished, the phosphorus
becomes luminous at a lower temperature; and
reciprocally, if the pressure be increased, the temperature must be elevated proportionally to make
the phosphorus shine. The introduction of some
foreign gas, such as hydrogen, nitrogen, or carbonic acid, into the mixture, has the same effect
upon the luminosity of phosphorus in air or oxygen, as if the pressure of the latter were diminished—a remarkable phenomenon observed by
M. Bellani. This is the reason phosphorus shines
at a lower temperature in the air than in pure
oxygen gas.

Thenard has made a curious experiment: he
shows that nitrogen, hydrogen, or carbonic acid,
which have remained for five or six hours in contact with phosphorus, and have then been separated from it, become luminous when a few bub-

D

bles of air or oxygen are passed into the gas experimented with. Carburetted hydrogen did not give the same result. This deserves to be examined anew, for I have shown in my 'Recherches nouvelles sur le Phosphore,' that phosphorus is not volatile at the ordinary temperature of the atmosphere.

According to Fischer, phosphorus is luminous in the atmosphere at any temperature above zero (freezing-point); it is even luminous at 6° (centigrade), but does not then appear covered with vapours. At a lower temperature its light disappears completely. In the barometric vacuum no light is produced by phosphorus. (See Fischer, "On the Light of Phosphorus," in 'Journal für praktische Chemie,' t. xxxv. p. 342.)

It is not true that phosphorus becomes luminous in carbonic acid, carbonic oxide, oxide of nitrogen, and cyanogen, as some have asserted. When such appears to happen, the gases are found to contain small quantities of air. The smallest quantity of air is indeed sufficient to occasion a production of light in these circumstances.

A solution of phosphorus in spirit of wine is luminous when dropped into water; the light is only perceived where the drops fall into the liquid. One part of phosphorus communicates this property to 600,000 parts of spirit of wine.

Water in which phosphorus is preserved, becomes luminous in the dark after a certain time.

When phosphuret of calcium is thrown into water, a decomposition takes place, and phosphuretted hydrogen gas is evolved. Each bubble of this gas, as it comes in contact with the atmosphere, takes fire spontaneously, and throws off a ring of white smoke. These pretty rings of smoke are luminous in the dark.

In the year 1851, M. Petrie discovered that the metal potassium is phosphorescent when exposed to the air, like phosphorus. ('Annuaire de Millon et Reiset,' 1851.) He covered the potassium with bees'-wax, and then cut it in two. Each segment remained luminous for about half an hour, the light being about one-tenth the intensity of that produced by a piece of phosphorus of the same size.

M. Linnemann published another note in 1859 (Journ. für prak. Chem., lxxv.), upon the phosphorence of potassium and sodium, showing that both these metals are luminous upon their freshly-cut surfaces. The light emitted by potassium is of a reddish tint, that of sodium greenish, according to this author. At 60° or 70° (centigrade), the light of sodium is quite as intense, if not more so, than that of phosphorus. I have had occasion to examine sodium whilst phosphorescent. Its light is very feeble at the ordinary temperature of the atmosphere, and ceases when the newly-exposed surfaces of the metal are covered with a layer of oxide (soda). The luminosity lasts for a few minutes, and increases in brilliancy as the temperature rises.

Potassium also becomes incandescent when employed in the preparation of boron, as must have been remarked by any chemist who has prepared this metalloid. For this purpose vitrified boracic acid, in fine powder, and potassium are heated in a metallic tube.

In fact, light, often accompanied by heat, is evolved, wherever chemical action is very intense. For instance, when sulphur and lead are melted together, light is produced whilst the combination of these two substances takes place. The same is remarked when phosphorus and iodine act upon each other: the experiment is very striking, and occurs when small quantities of phosphorus are covered over with iodine, at the ordinary temperature of the atmosphere. In a short time the whole takes fire spontaneously. I have seen the same occur when a crystal of nitrate of copper was enveloped in a thin sheet of tin.

When arsenic or antimony are thrown into chlorine gas at the ordinary temperature, the metals (which must be in fine powder) burst into flame while combining with the chlorine. When caustic baryta is placed in a capsule and concentrated sulphuric acid poured upon it, the baryta becomes incandescent.

If a drop of water fall into a bottle of anhydrous

sulphuric acid, a flash of light, accompanied by a slight explosion, is immediately remarked. If small pieces of cork happen to fall upon melted chlorate of potash, so frequently used to obtain oxygen gas, a flash of light appears; the gas is at first rapidly evolved from the retort, but in an instant an absorption takes place, the water is sucked up into it, and the apparatus broken. When chloride of amide (formerly called chloride of nitrogen) explodes, much light is evolved: the preparation and explosion of this substance are exceedingly dangerous. Potassium takes fire upon water, and burns with a purple flame; I have also seen sodium shoot out flashes of yellow light in the same circumstances. Nitric acid decomposes oil of turpentine, producing a great flame. When great quantities of lime are slacked in a dark place, not only heat but *light* is emitted, as was formerly observed by Pelletier.* Also, in a dangerous experiment made by myself, when sodium

---

* There are substances called *Kacodyles*, (one of which is formed when acetate of potash and arsenious acid are distilled together,) which take fire spontaneously when they come in contact with atmospheric air: they are liquid, and possess a nauseous odour. Homberg's *pyrophorus*, which takes fire in the air, is an example of intense chemical action with production of heat and light. It is prepared by calcining alum with organic matters and cooling the mixture slowly. It must not be mistaken for Homberg's *phosphorus*, which is melted chloride of calcium, and which, as we have seen, becomes luminous when submitted to rapid friction, or when struck with a hard body.

is dropped into concentrated sulphuric acid, a flash of light, distinctly visible in the daytime, is emitted, and red sulphuret of sodium is formed.

These are all instances of intense chemical action, accompanied by evolution of light.

CHAPTER IV.

PHOSPHORESCENCE OF GASES, AND ELECTRIC PHOSPHORESCENCE.

THE phosphorescence of gases is quite a new discovery, dating from the year 1859. It is extremely probable that many gases are phosphorescent after insolation, when large quantities of them are submitted to the action of the sun's rays. We shall see in the following chapter that the air probably is so, and also that meteoric stones leave phosphorescent streaks in the atmosphere.

We have already noticed, that substances which are not phosphorescent after insolation may become so after they have undergone the influence of an electric discharge. In February, 1859, M. Edmond Becquerel communicated to the *Academy of Sciences* at Paris a discovery made by M. Ruhmkorff on rarefied air, and worked out afterwards by the former.

M. Ruhmkorff remarked that certain rarefied gases, shut up in glass tubes, remained phosphorescent for some seconds after an electric discharge had been passed through the tubes and gases. Hydrogen, sulphuretted hydrogen, chlorine, protoxide of nitrogen, showed a feeble light for a few seconds after being submitted to an electric discharge or a current of induction. With oxygen a similar effect is observed. Rarefied oxygen, enclosed in a serpentine apparatus composed of a series of glass globes united by bent tubes (fig. 4), in which are soldered platinum wires to

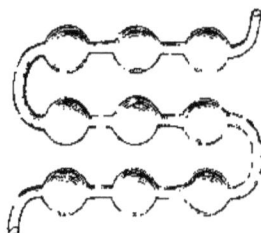

Fig. 4.

conduct the discharge, is submitted to the action of a powerful induction machine or common electric battery. When the current is suddenly cut off, the entire tube shines with a yellowish light, which persists for some seconds and then gradually disappears. The experiment must of course be made in a dark room.

Sulphurous acid gas sometimes shows a similar effect. M. Ed. Becquerel has not been able to observe phenomena of phosphorescence in any of these gases after insolation or after exposing them to the electric light, though it is probable such does exist.

In the above experiment, when other gases are mixed with the rarefied oxygen, the effect is somewhat increased, probably because a certain amount of chemical action is set up.

Our readers know, that by passing the discharges of a Ruhmkorff's induction apparatus through a glass globe in which the air has been highly rarefied, a beautiful luminous phenomenon occurs and persists as long as the induction apparatus continues to work. When the vapour of some volatile substance, such as alcohol, essence of turpentine, naphtha, bichloride of tin, etc., is mixed with the air in the glass globe, and a vacuum then produced by the pneumatic machine, the luminous phenomenon is still more beautiful. The light forms a series of concentric arches separated by dark stratifications; its colour and form remind us of the Aurora Borealis; and, indeed, some have looked upon this experiment as the production of an artificial aurora, the vacuum of the glass globe (which is never a perfect void) representing the rarefied air in the higher regions of the atmosphere, where the Aurora Borealis occurs.

The light thus produced, to be seen to advantage, must be viewed in a dark room, but it is faintly visible in full daylight.

These striking experiments were made some few years ago by Professor Quet, and have excited general admiration wherever they have been seen. Now it has occurred to M. Ed. Becquerel, to enclose certain phosphorescent substances, such as sulphide of barium, etc., in glass tubes, in which a vacuum has been produced by the air-pump, and to submit them in these circumstances to the action of M. Ruhmkorff's apparatus.

We have already seen that sulphide of barium, of strontium, of calcium, diamonds, chalk, etc., acquire phosphorescence when submitted to an electric discharge in the air; as soon as the discharge has passed, they glow with phosphoric light of short duration, just as if they had been exposed to the sun, or as if they had been heated; for these substances are phosphorescent by light, by heat, and by electricity. But when they are submitted to the rapid series of discharges of the induction apparatus in *highly-rarefied air*, that is, in the void produced by the air-pump, the effect is very striking. The substances named glow continuously with a vivid phosphoric light, so long as the discharges continue to pass.*

In these experiments it has been observed, that the glass of the tubes becomes slightly phosphorescent at the same time as the sulphides.

Quet made known in 1853 a very curious pro-

* See Ed. Becquerel, Ann. de Chim. lv. p. 92 *et seq.*

duction of light that takes place when water is being decomposed by the electric current, after being rendered a good conductor of electricity by the addition of sulphuric acid or potassa. When forty Bunsen's elements are employed, the water is rapidly decomposed, and its temperature considerably raised, but a moment comes when the platinum wires plunging into the liquid *become suddenly luminous, and the decomposition of the water ceases as suddenly.* In this case, when the water is acidified with sulphuric acid, the light given out by the positive pole is red, whilst that emitted by the negative pole is violet. The light seems to encase the wires and to repel the water, so as to prevent its contact with the metal.

CHAPTER V.

METEOROLOGICAL PHOSPHORESCENCE.

NUMEROUS observations leave no space for doubt regarding *phosphorescence of the drops of rain* in certain storms. The phosphoric light produced in these circumstances shows itself upon the coats of travellers, or on the borders of their hats, etc. This phenomenon astonished M. de Saussure whilst travelling on the summit of the Breven; whenever he lifted his hand, he felt a sort of creeping sensation in the fingers, and in a short time an electric spark was drawn from a golden button affixed to the hat of his companion, M. Jalabert. The storm roared in the clouds around him.

A somewhat similar phenomenon occurred to Dr. Kane, the intrepid Arctic explorer, which, for certain reasons, we shall speak of in a future chapter.

On the 25th of January, 1822, during a heavy shower of snow, M. de Thielaw, on his route to

Freyburg, observed the *branches of the trees glow-ing with a bluish light.*

François Arago has collected many instances of luminous rain, among which are the following:—

On the 3rd of June, 1731, Hallai, an ecclesiastic of Lessay, near Constance, states that he saw, in the evening, during a thunderstorm, *rain which fell like drops of red-hot liquid metal.*

In 1761, Bergman, the celebrated Swedish chemist, wrote to the Royal Society of London that he had observed on two occasions, towards evening, and when no thunder was heard, *rain which sparkled as it touched the ground*, making the latter appear as if covered with waves of fire.

On the 3rd of May, 1768, near Arnay-le-Duc, M. Pasumot was overtaken on an open plain by a violent storm. The rain-water collected abundantly on the border of his hat; and when he stooped his head to let it flow off, he observed that, in its fall, encountering that which fell from the clouds, at about twenty inches from the ground, *sparks were emitted* between the two portions of liquid.

On the 28th of October, 1772, on his way from Brignai to Lyons, the Abbé Bertholon was caught in a storm at five o'clock in the morning. Rain and hail fell heavily. The *drops of rain and the hailstones* which struck against the metallic parts of the mounting of his horse's trappings, *emitted jets of light.*

A friend of Howard, the meteorologist, on his way from London to Bow on the 19th of May, 1809, during a violent storm, distinctly saw the *drops of rain emit light* when they struck the ground.

On the 25th of January, 1822, the miners of Freyburg informed Lampadius that *the sleet* which fell during a storm *emitted light* when it struck the earth.

There are other similar instances; these phenomena are evidently closely allied to electricity.

Waterspouts (called by the French *les trombes*), according to Peltier and more recent observers, are sometimes observed to be *luminous* when they happen in the night.

A case of *luminous meteoric dust* is also on record:—During the eruption of Vesuvius which took place in 1794, a shower of extremely fine dust fell in Naples and its environs. It emitted light which, though pale, was distinctly visible at night. An English gentleman, who happened to be in a boat near *Torre del Greco* about this time, observed that his hat, those of the boatmen, and parts of the sails where the dust had lodged, shed around a sensible luminosity.

Shooting stars, or *meteoric stones,* leave after them in the heavens a *phosphoric stream of light,* which often persists for a considerable time after their passage (fig. 5). In his voyage round the world, the Admiral de Krusenstern saw one of these Aerolites leave behind it in the sky a

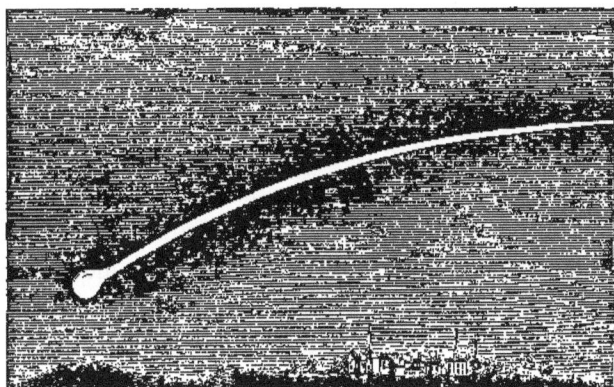

Fig. 5.

phosphorescent streak which persisted for a whole hour, without sensibly changing its place. (See Humboldt: 'Cosmos,' vol. i.) Phosphorescent streaks left behind Aerolites not unfrequently remain visible for about a minute.

We cannot do more than mention here the lightning flash,* the Aurora Borealis, the Zodiacal light, the fire of St. Elmo, the light of fixed

* On the various kinds of lightning, see Arago, "Notice sur le Tonnerre," in his 'Œuvres,' or in the Ann. du Bureau des Longitudes, for the year 1838 ; Phipson, in the ' Comptes-Rendus' of the Academy of Sciences of Paris, April 13, 1857 ; and Du Moucel's *brochure*, 'Sur le Tonnerre et les Eclairs.' Paris : Hachette, 1857.

stars or suns, and the flame, all of which doubtless belong to our present subject.

Lord Napier observed the fire of St. Elmo in
the Mediterranean during a fearful thunderstorm.
As he was retiring to rest, a cry from those aloft
of " St. Elmo and St. Anne !" induced him to go
on deck. The maintop-gallant-mast head was
completely enveloped in a blaze of pale, phosphoric
light, and the other mast-heads presented a similar
appearance. The phenomenon lasted for eight or
ten minutes, and then became gradually fainter.
All other descriptions of this electrical phenomenon
coincide perfectly with the above.

The Zodiacal light, when seen under the tropics,
often shines with a brilliancy equal to that of the
Milky Way in Sagittarius. In our Northern climates, it is only observed shooting up towards
the Pleiades in the beginning of spring, after the
evening twilight, in the western part of the sky ;
and at the close of autumn, before the dawn of
day, above the eastern horizon.

Some philosophers have asserted that the sun's
light is an effect of combustion, like the flame of a
common candle; but, from a comparison of the rela-
tive intensities of solar, lunar, and artificial light,
as determined by Euler and Wollaston, it appears
that the rays of the sun have an illuminating power
equal to that of 14,000 candles at the distance of a
foot, or of 3,500,000,000,000,000,000,000,000,000

candles at the distance of 95,000,000 of miles, which is our distance from the sun. Hence it follows, that the amount of light which flows from the sun could scarcely be produced by the daily consumption of 700 globes of tallow, each equal to the Earth in magnitude.

It does not follow that because planetary bodies shine principally by borrowed light, that they do not possess also a certain amount of *phosphoric luminosity*. Some modern philosophers are inclined to believe that our earth itself has a peculiar phosphoric light of its own :—

"The extraordinary *lightness of whole nights* in the year 1831," says Alex. von Humboldt, "during which small print might be read at midnight in the latitudes of Italy and the north of Germany, is a fact directly at variance with all that we know according to the most recent and accurate researches on the crepuscular theory and the height of the atmosphere." (Cosmos, i. 133, 134.) And, again in the same beautiful work (p. 197), when speaking of the Aurora Borealis :—"This beautiful phenomenon derives the greater part of its importance from the fact that the earth becomes *self-luminous*, and that as a planet, besides the light which it receives from the central body the Sun, it shows itself capable in itself of developing light." The intensity of the light thus diffused is often superior to that shed by the moon in her

E

first quarter. "Occasionally, as on the 7th January, 1831, printed characters could be read without difficulty." Indeed, the intensity of the Northern or Polar light is sometimes very great; and Lowenörn assures us, that on the 29th June, 1786, he recognized the coruscation (trembling motion) of the Aurora Borealis in bright sunshine.

Some splendid manifestations of the *Aurora Australis,* or Southern light, were witnessed by Captain J. Ross in his voyage to the South Pole.* These Southern lights have often been seen in England by Dalton, and Northern lights have been witnessed in the Southern hemisphere—14th January, 1831—as far as 45 degrees latitude.

It has been shown, by many experiments, that the electric light, whatever be its source, does not show any signs of polarization. Now this has been lately proved to be the case with the light of the Aurora, which shows no polarization, according to Mr. Rankine; but when it is viewed reflected from the surface of a river, polarization is detected, —which proves that in the former case the want of polarization is not owing to the weakness of the Aurora observed.

A curious phenomenon was noted by Admiral Wrangel, when he was on the Siberian coast of the Polar Sea. He observed, that during an Aurora

* The *Aurora Australis* was seen for the first time by Captain Cook in his first voyage, and again in his second voyage.

Borealis, certain portions of the heavens which were not illuminated, lit up and continued luminous whenever a shooting star passed over them.

M. Colla, formerly director of the Observatory of Parma, has often observed, since 1825, a singular light in the northern sky, like a zone of 10 or 12 degrees, parallel to the horizon, often of a yellow colour, and most intense in the direction of the magnetic meridian. He considers it to be "the permanent element of the Aurora Borealis."

That portion of the planet Venus which is not illuminated by the sun often shines with a *phosphorescent light of its own.*

Towards the latter end of June, 1861, the earth passed through a region of the heavens, then occupied by a portion of the great comet of that year. On this occasion Mr. Hind, Mr. Lowe, and others, observed a peculiar phosphoric glare in the atmosphere. It was remarked by many persons as something unusual.

That portion of the moon which is not illuminated by the solar rays shines with a grey light of its own, called by the French *lumière cendrée.* This is generally attributed to the light thrown upon our satellite by the illuminated portions of the earth; but it may be that the moon possesses phosphorescent qualities like other celestial bodies.

Doubtless other planets possess similar phosphorcscent properties :—"It is not improbable," says Humboldt, "that the Moon, Jupiter, and comets shine with an independent light besides the reflected, solar light, visible through the polariscope."

We cannot do better than quote the following passage also, by the same author :—

"Without speaking of the problematical, but yet ordinary mode in which the sky is illuminated when a low cloud may be seen to shine with an uninterrupted flickering light for many minutos together [see further on], we still meet with other instances of terrestrial development of light in our atmosphere. In this category we may reckon the celebrated *luminous mists* seen in 1783 and 1831 ; the steady luminous appearance exhibited without any flickering in great clouds observed by Rozier and Beccaria; and, lastly, as Arago well remarks, *the faint diffused light which guides the steps of the traveller in cloudy, starless and moonless nights, in autumn and winter, even when there is no snow on the ground.*"

Indeed, any attentive observer of Nature may assure himself that in the darkest nights of winter, at the hour of midnight, when the influence of solar light is altogether withdrawn from the atmosphere, and in the absence of moonlight, a sufficient quantity of light is always diffused to render objects around us faintly visible, and to enable us to walk without hesitation in any open country.

Let the heavens be overcast, let the stars be hidden by an unbroken mass of clouds, and still a sufficiency of light will be diffused in the open country to prevent the difficulty and inconvenience which would attend any attempt to walk in a dark cave, or in an apartment the shutters of which are closed.

It appears to me that the atmosphere and the clouds themselves act in these cases like the *phosphori* spoken of in a previous chapter. Being exposed to the light of the sun the whole day long, it is very probable that they emit a phosphorescent light like the Bologna stone, for instance, when the Sun's rays are withdrawn from them; and moreover, that this phosphorescence may, in certain circumstances, assume an extraordinary intensity, as we shall see by some of the following examples.

Here are the accounts by Rozier and Beccaria, alluded to above :—

Rozier states that, being at Beziers, in France, on the 15th August, 1781, at a quarter before eight in the evening, the sun having gone down and the sky overcast, thunder was heard. At five minutes past eight, the storm having attained its height, Rozier observed a luminous point above the brow of a hill fronting his house; this point gradually augmented in magnitude until it assumed the form and appearance of a phosphoric zone subtending at his eye an angle of about 60 degrees, measured horizontally, and having the apparent height of a few feet; above this was a dark band, and then again another zone of light. These *luminous zones of cloud* were nearer the earth than the storm clouds, and their brilliancy lasted about a quarter of an hour.

Beccaria assures us that the clouds over his observatory at Turin frequently *shed in all directions a strong reddish light*, which was sometimes so intense as to enable him to read ordinary print. This nocturnal luminosity was especially observed in winter, between successive falls of snow.

When General Sabine and his crew were lying at anchor at Loch Scavig, in the Isle of Skye, he observed a cloud which constantly enveloped the summit of one of the naked and lofty mountains which surround that island. This cloud which had been formed by the vapour precipitated near the mountains after having been brought by the constant west winds from the Atlantic, *was self-luminous* at night, not occasionally, but permanently. He saw frequently issue from it jets of light, and convinced himself that this phenomenon had nothing whatever to do with the Aurora Borealis.

We may add to these an observation of Nichol-

son, who states that during a storm on the 30th
July, 1797, at about five in the morning, certain
clouds were observed to shine first with a red, and
afterwards with a blue, light.

De Luc affirms also, that one winter's night, in
the neighbourhood of London, he observed a *luminous cloud* extending east and west across the
southern meridian of the place, about 30 or 40
degrees from the zenith. The atmosphere was
clear but not cold, and "there were no signs of
electricity."*

One of the most authentic and curious observations of *luminous fogs* was lately communicated in
a letter to M. Elie de Beaumont by M. L. F.
Wartmann,† of Geneva. The strange phenomenon
was observed during nine successive foggy nights,
from the 18th to the 26th of November, 1859. The
moon being new, was invisible and absent from
the heavens of Geneva. But a vast fog, not damp
enough to wet the earth, but so opaque as to
render invisible the borders of the river Leman
and the mount Salèse, hovered permanently over
Geneva and its environs. This fog *diffused so
much phosphoric light,* that M. Wartmann could
easily distinguish books, etc., upon his table, with-

* For more ample details on some of these phenomena, see
Beccaria, 'Dell' Elettricismo terrestre atmosferico;' Deluc, 'Idées
sur la Météorologie,' and Arago, in the 'Annuaire' for 1838.

† Comptes-Rendus of the Academy of Sciences, Paris, 25 December, 1859.

out having recourse to any other light. Moreover, he questioned a person who had gone on foot from Geneva to Annemasso, in Savoie, on the 22nd of November: he had started at half-past ten at night, and declared that he saw his way the whole distance as if it had been a moonlight night. M. Auguste de la Rive was, that same night, at some distance from Geneva, and was also surprised at the distinctness with which he saw his road and the objects around him.

The celebrated *dry fog* of 1783 was described by M. Verdeil, a physician of Lausanne, as having diffused at night, a *luminosity sufficiently intense* to render distant objects visible, and this light was equally spread in all directions. It resembled the light of the moon seen through the clouds.

This dry fog, in which objects could be seen at night at a distance of 600 feet, lasted a whole month; it made its appearance nearly at the same time in many distant places, spreading from the north of Africa to Sweden; it was likewise observed over a great portion of North America, but was not seen to spread over the sea. It appeared to reach higher than the summits of the highest mountains, and neither winds nor rain had any power to disperse it. In Europe this fog exhaled a disagreeable odour, was remarkably dry, did not affect the hygrometer, and possessed the remarkable *phosphoric quality* I mentioned above.

Many philosophers thought that, at this period, the earth was bathed in the tail of a comet.

But in 1831 another dry fog exactly similar was observed; it did not spread so far as that of 1783, and as it did not cover the whole of Europe, it was easy to perceive that no comet was present to cause its production.

The origin of these dry, luminous mists, is yet a mystery. It may, however, be noted that in 1783 Calabria was visited by a terrible earthquake which destroyed 40,000 inhabitants; Mount Hecla, in Iceland, broke out in one of its most remarkable eruptions, and volcanic rocks were seen to emerge from the sea, etc.

It is said that a periodical dry fog, which does not spread over the sea, visits the eastern coasts of Africa with the disastrous wind called *the parmatan;* but whether it is luminous or not I cannot say.*

But one of the most curious phenomena ever witnessed was doubtless that described as having been seen by General Sabine and Captain James Ross in their first northern expedition. Being in the Greenland seas during the period of darkness, they were called up by the officers on deck to observe an extraordinary appearance. Ahead of the vessel, and lying precisely in her course, appeared

* On a dry fog observed at London, see my note in the 'Comptes-Rendus,' Paris, 1861.

a stationary light resting on the water, and rising to a considerable elevation. Every other part of the heavens and the horizon all around the ship were in utter darkness. (*Vide Frontispiece.*) As there was no known danger in this phenomenon, the course of the vessel was not altered; and when the ship entered the region of this light, the officers and crew looked on with the liveliest interest. *The whole vessel was illuminated;* the most elevated parts of the masts and sails, and the minutest portions of the rigging, became visible.

The extent of this *luminous atmosphere* might have been about 450 yards. When the bow of the ship emerged from it, it seemed as if the vessel were suddenly plunged in darkness. There was no gradual decrease of illumination. The ship was already at a considerable distance from the luminous region when it appeared still visible as a stationary light astern.

Many persons would look upon this curious phenomenon as an intensely *phosphorescent mist.* Several observations tend to prove that in these northern latitudes the density, and often the dryness, of the air, contribute much to the intensity of luminous apparitions, especially those which appear to depend upon electrical actions. The above is the account of this phenomenon as related in F. Arago's 'Notice sur le Tonnerre.'

The following is however that of General Sabine himself, which he has kindly given me in a letter:—"Before the ship entered into the Auroral light, the Aurora as seen from the ship appeared as an *arch*, formed partly of an uniform yellowish light, and partly of vertical or nearly vertical streamers proceeding out of the luminous arch upwards. The centre of the arch was not far distant from the zenith, and the legs descended towards east and west points. We were opposite to one of the legs, and sailing towards it till we entered it. We were sensible of having entered it, by no longer seeing it as a distant appearance, and by the moment of our entrance into it being marked by a generally diffused light, enabling those on deck to see distinctly men on the foretopsail yard, who we could not see previously." The ship was sailing southward, and entered the western leg of the luminous arch.

At the meeting of the Literary and Philosophical Society of Manchester on January 31, 1861, Mr. Baxendell stated that many of the fogs observed during that winter were luminous. Mr. Crosse and other observers have found fogs to be highly electrical.

I will place here a passage from a Brussels correspondent, who writes in 1860 :—

" On looking out of my bedroom window at two o'clock on the morning of the 25th January, I perceived *the sky to be very light,* insomuch that I could discern the buildings and other objects. The wind blew in fresh gales from W.S.W., barometer below 'tempest,' and thermometer about 41°. I have only once seen this luminosity before."

Loch Scavig appears destined to become celebrated for luminous phenomena. Besides the phosphorescent cloud seen there by General Sabine, my friend Mr. T. K. Edwards tells me of another curious case of a luminous meteor seen in the same locality. It was in the month of September, 1852 or 1853, and the phenomenon was observed about eight o'clock in the evening. He was being rowed by four stout men from Torrin, in the Isle of Skye, to one of the opposite shores. On entering Loch Scavig the boat containing Mr. Edwards, his friend Mr. Raymond, four boatmen, and a guide, steered across the little bay situated on their right, when a light was distinctly seen at a great distance to the seaward. At first it appeared like the light from the cabin window of a steamboat being near to the surface of the water, and moving with great rapidity towards them. The four men at the oars noticed it with evident alarm, and spoke hurriedly to each other in Gaelic. When the guide was asked what they were talking about, he answered, " About yon light ; it's no canny thing, neither." The rapidity with which the light moved, and its proximity to the boat after a few seconds had elapsed, fully convinced every one that it belonged to no boat; besides, as the guide remarked, "no bird could fly so quick." It appears that this phenomenon, which I believe to have been globular lightning, is not unprecedented in the neighbourhood of Loch Scavig; for though the four oarsmen had never witnessed it before, they had heard it spoken of on the land as betokening evil, and were so much afraid of it that they pulled the boat along most lustily. The light curved off and was soon lost to sight, having been in view and indeed very near to the boat, for about two minutes. The next day was extremely sultry.

This kind of travelling light reminds us of some descriptions of Will-o'-the-Wisp; but besides being seen over the sea, its resemblance to the light of a ship (though it is quite evident no ship or boat carried it), and the extreme sultriness of the next day, makes me think that it is more probably allied to those curious cases *globular lightning*. Our travellers in the boat may not have noticed the sultriness of the air whilst on the water, but only remarked it the next day, and the men at the oars might have heard of the disastrous effects of globular lightning.

A similar light, but a fixed one, was observed by Maffei, in 1713, and the phenomenon recorded by F. Arago:—It was during a heavy shower of rain on the 10th of September. He took shelter in the Château de Fosdinovo (in the province of Massa-Carrara, in Italy), and was standing at a window on the ground-floor, when a bright light appeared upon the pavement. This light, which in colour appeared to be white and blue, was very unsteady or vibrating, but had no progressive motion; it disappeared as it had made its appearance, that is, suddenly, but not until it had acquired a great size. The moment it vanished, Maffei felt a slight itching on his shoulder, a little plaster fell from the ceiling of the room, and a cracking noise was heard, very different from that of thunder.

A flame which rose from the ground and maintained itself at three feet from the floor of the church St. Michel, at Dijon, was noticed by the Abbé Richard, on the 2nd July, 1750. This flame rose afterwards to twelve or fifteen feet in height, increased considerably in volume, and disappeared near the organ of the church with a loud explosion.

These phenomena are evidently allied to atmospheric electricity, as they manifest themselves during storms; but F. Arago distinctly states ('Annuaire,' 1838, p. 371): "it appears that great luminous meteors, similar to lightning in their nature, show themselves sometimes at the surface of the globe, even *when the sky does not appear stormy."* Such was the case on the 4th November, 1749, in 42° 48' latitude north and $11\frac{1}{3}$° longitude west of Paris: a few minutes before mid-day, and in *perfectly serene weather,* a large bluish globe of fire rolled up to the ship the 'Montague,' and exploded, shattering one of the masts. This globe of fire appeared as large as a millstone. A strong "sulphurous" odour was perceived in the ship for some time afterwards. The light described to me by Mr. Edwards appears to have been some such phenomenon, and had he and his companions seen the end of it, the fears of the boatmen might have been realized.

Detailed accounts of similar *electric* meteors may be foreign to the subject of the present work, though electricity plays, doubtless, its part in all phosphoric phenomena; but I have endeavoured in these pages to notice, however briefly, *every known source of terrestrial light,* for it is not in our power, in the present state of science, to restrict phosphorescence to a limited number of phenomena.

I must now say a few words on that beautiful and mysterious production of light known as the *Will-o'-the-Wisp* or *Ignis fatuus (feux follets* of the French).

This phenomenon is generally attributed, by chemists of the present day, to the spontaneous inflammation of phosphuretted hydrogen gas. It

is well known that one of the gaseous compounds
of phosphorus and hydrogen takes fire as soon
as it comes in contact with atmospheric air; and
it is supposed that in certain circumstances the
putrefaction of animal matters, containing phosphorus and sulphur, besides the four elements
carbon, hydrogen, nitrogen, oxygen, and phosphate of lime, is accompanied by a production of
this phosphuretted hydrogen gas. *Will-o'-the-
Wisp* is observed in boggy lands, and where it is
seen some animal or perhaps an unlucky traveller
has been swallowed up in the mire.

The "*corpse-candle*" of the Welsh, which flickers
over churchyards, is attributed to the above cause,
and the same may be said of that mysterious production of light which occasionally takes place in
dissecting rooms.

But no chemical experiment, made with organic
matters, has yet been brought forward to prove
the production of phosphuretted hydrogen with evolution of light *by submitting these matters to the
process of putrefaction.* Indeed, I have shown, as
will be stated in a future chapter, that the phosphorescence of dead fish does not appear to depend
upon the presence of the chemical element phosphorus.

If, however, it were placed beyond doubt that
the phenomenon of the *Will-o'-the-Wisp* or *Ignis
fatuus,* depended upon the production and spon-

taneous inflammation of phosphuretted hydrogen,
it could not be classed among phenomena of
phosphorescence any more than the flames of
certain *fire-springs* in the East, which are owing
to the combustion of carburetted hydrogen or
naphtha, and some of which, like the famous
Lycian Chimæra, in Asia Minor, have been burning for several thousand years.

From some very interesting arguments brought
forward in the last edition (in one volume) of
Kirby and Spence's ' Introduction to Entomology,'
it would appear probable that some cases of *ignis
fatuus* might be attributed to certain luminous
insects not yet known, which hover in clusters
over marshy ground.   These insects seem to belong to the genus *Tipula* (Gnat, "Daddy-Long-
legs," etc.), if we are to judge from the hovering
appearance of the light. Thus Dr. Derham, in the
' Philosophical Transactions' for 1729, describes
an *ignis fatuus*, seen by himself, as flitting about
a thistle.

Dr. Derham got within two or three yards of
another *ignis fatuus*, in spite of the boggy nature
of the soil.   He states, however, that it appeared
like *a complete body of light without any division*,
so that he was sure *it could not be occasioned by
insects*.

At the same time, it is evident that no insects
could produce the phenomenon described by Dr.

F

Weissenborn, in 1818, where the light travelled over a distance of half a mile in less than a second. (Mag. of Nat. Hist. N.S. i. 553.)

From these facts, it appears probable to modern philosophers, that there are two kinds of *ignes fatui*, the one referable to spontaneously inflammable gas, the other to luminous insects.

If phosphuretted hydrogen, or any other spontaneously combustible gas or liquid, caught fire upon a marsh where carburetted hydrogen (marsh gas) is constantly evolved, the latter would inflame also.*

In the valley of Gorbitz, Mr. Blesson discovered a light emanating from marshy ground. Remaining for some days near the place, in order to study the phenomenon as closely as possible, he found it was owing to an ignited gas, the faint flame of which was invisible during the day, but became gradually visible in the evening. The gas appears to have been carburetted hydrogen, or marsh gas. As he approached it, the flame receded, but he eventually succeeded in lighting a piece of paper by it.

According to some authors, *Will-o'-the-Wisp* may be seen at all seasons of the year; but a great

* At Wigmore, in Herefordshire, and other places in England, carburetted hydrogen used to be so abundant in the ground that it was employed for lighting and cooking in the houses, as we learn from travellers is a common practice in China.

number of persons I have questioned upon the subject all agree in stating that they have noticed it in autumn, or towards the end of autumn, and the beginning of November. It cannot be termed a common phenomenon, as many distinguished naturalists have never been able to observe it; it is not unfrequent in the north of Germany, and is often witnessed in the peat districts around Port Carlisle, in the Lowlands of Scotland, and in the swampy parts of the South of England, etc. It was seen by Mr. Warltire, in a very curious form, on the road to Bromsgrove, about five miles from Birmingham, as noted by Dr. Priestley in the Appendix to his third volume of 'Experiments and Observations on Air.' The time of observation was the 12th December, 1776, before daylight.

Some countries are very remarkable for this curious phenomenon; for instance, the neighbourhood of Bologna, in Italy, and some parts of Spain and Ethiopia. According to an account left us by M. Beccari, an intelligent gentleman travelling in the evening, between eight and nine o'clock, in a mountainous road about ten miles South of Bologna, perceived a light which shone very strangely upon some stones on the banks of the Rio Verde. It appeared as a parallelopiped of light, about a foot in length, and two feet above the stones. Its light was so strong that he could plainly see by it part of a neighbouring hedge and the water of the river. On examining it a little nearer, he was surprised to find that the light became paler, and when he came to the place itself, it quite vanished. *No smell* or other mark of fire *was observed* at the place where this light shone.

Another gentleman informed M. Beccari that he had seen the same light five or six different times, in spring and autumn, always of the same shape, and in the very same place.

Dr. Shaw, in his 'Travels to the Holy Land,' states that an *ignis fatuus* appeared to him in the valley of Mount Ephraim, and attended him and his company for more than an hour. Sometimes it appeared globular, at others *it spread to such a degree as to involve the whole company in a pale, inoffensive light;* then contracted itself, and suddenly disappeared, but in less than a minute it would appear again; sometimes running swiftly along, it would expand itself at certain intervals over two or three acres of the adjacent mountains.

Dr. Priestley has given an account of what some look upon to have been an artificial *Will-o'-the-wisp.* A gentleman, who had been making electrical experiments for a whole afternoon in a small room, on going out of it, *observed a flame following him at some little distance.* In this case, however, there seems to have been a difference between the artificial *ignis fatuus* and that met with in nature, for the flame *followed* the gentleman as

he went out of the room, but the natural phenomenon generally recedes as we approach it.

It is a common practice, in chemical lectures, to
imitate the *Will-o'-the-wisp* by throwing fragments
of *phosphuret of calcium* into water, when flames
arise, owing to the spontaneous combustion of
phosphuretted hydrogen gas. But the imitation
is very bad indeed, and can hardly be said to
resemble the mysterious natural phenomenon,
much less explain it. For my own part, I think
the *ignis fatuus* to be sometimes the light from a
burning gas, which light is invisible in the daytime, and at other times to be connected with
those curious cases of luminous mists mentioned
above, and in which electricity doubtless plays an
important part.

The luminous appearances known in Scotland
as *Elf-candles* belong either to this category of
phenomena, or to that which will be treated of in
a future chapter of this Work.

CHAPTER VI.

DURATION, INTENSITY, AND COLOUR OF PHOS-PHORIC LIGHT IN MINERAL BODIES.

THE duration, the intensity, and the colour of phosphoric light produced by mineral substances, *depend upon the nature of the phosphorescent body.*

I shall mention only a few examples of colour. The most general tint of light is that seen in the glow-worm and other phosphorescent animals, of which we shall speak hereafter; it is a greenish-yellow light, at times approaching to whiteness. Some bodies however appear, during their phosphorescence, to emit light which differs a little from this as to its colour. Certain marbles and amber (succinum) give a phosphorescent light of a golden yellow; some specimens of fluor-spar, arseniate of lime, and chloride of calcium, emit a greenish light; other varieties of fluor-spar produce a bluish-violet radiation, and that which is called Chlorophane gives a green phosphorescence. Oriental garnet shines with a reddish phosphorescence, whilst Harmotome (a sort of zeolite) gives a greenish-yellow phosphorescence. Dolomite, Aragonite, and some specimens of diamond, shine with a brilliant, white, phosphoric light.

In the same manner, the colour of a flame depends upon the nature of the body that burns. Thus, carburetted hydrogen and sodium burn with a yellow flame, oxide of carbone with a blue flame, potassium and cyanogen with a purple flame, etc.

Pearsall, Brewster, Dessaignes, and Becquerel have studied this subject. It appears to me very evident that the same substance may slightly differ in the colour of its phosphorescence, according to the manner in which the latter is prepared or excited.

Concerning colours and tints, we should, in general, be careful in admitting them too exclusively, for there are scarcely two persons who will entirely agree upon the denomination of any tint that is not one of the striking colours of the solar spectrum.

CHAPTER VII.

INVISIBLE PHOSPHORESCENCE.

I HAVE given the name of Invisible Phosphorescence to some curious phenomena discovered by
my ingenious friend M. Nièpce de St. Victor,
who communicated them to me with much kind-
ness before they were published. During latter
years he has, however, addressed to the Academy
of Sciences, at Paris, a number of notes and papers,
in which his experiments are detailed.* The basis
of them all was the following interesting observation :—

M. Nièpce discovered that if an engraving be
exposed for some time to the sunlight, and then
taken into a dark room, and placed upon a sheet
of photographic paper prepared with chloride of
silver, an impression of the engraving is produced
in a very short space of time upon the paper. This
experiment was immediately tried with a great
variety of substances, such as white porcelain with

* 'Comptes-Rendus' from 1857 to the present time.

black figures, wood, linen, cardboard, marble, etc., and always with the same result. In every experiment, an impression was left on the photographic paper in the dark, and this experiment was clearly proved to be owing to the *action of light alone;* no chemical agent whatever to which such a phenomenon might be attributed, entered into these experiments.

It was soon perceived that certain substances seemed to possess, as it were, "a greater affinity for light than others," and, as M. Nièpce used to say, "seemed to become, during their exposure to the sun, more saturated with light than others" in the same space of time, and, consequently, acted with greater intensity on the photographic paper in the dark.

A step more, and my friend had actually "bottled up light," to use his own expression. A sheet of cardboard, imbibed with a solution of tartaric acid or a salt of uranium, was rolled into a cylinder and placed inside a tin tube, open at one end, so as to line it. The tube was then exposed to the light, with its orifice towards the sun. After a certain time had elapsed, from a quarter of an hour to about an hour, the orifice of the tube was hermetically sealed up. If such a tube be taken into a dark room, opened, and its orifice placed upon a sheet of photographic paper, in a very short time the impression of this orifice is left upon the

paper. I have seen tubes of this kind that had
been prepared and corked up for a week, a fort-
night, and some that had even been closed up for
months, after their exposure to light, and all left
the impression of their orifices upon the photographic paper, as if the paper in the tube had
been acted upon only a few seconds before. But
the impression is not so intense when the tube
has been kept closed for a long period of time.

These effects are owing probably to a phenomenon of phosphorescence, and, if so, they prove
evidently that all bodies possess this property to
a greater or less extent, depending upon the
nature of the substance examined. Luminous
vibrations, which constitute phosphorescence, are
hereby shown to exist when we cannot perceive
them : their presence is made known by the photographic paper when the eye is not able to discern
them. These luminous vibrations persist, also, for
a period of time which is much longer than any
one would, at first, be inclined to suppose.

In a recent paper, M. Nièpce says :—" I have
repeated my former experiments of shutting up
light in tubes, employing in these experiments
cardboard imbibed with a salt of uranium, or with
tartaric acid. The results have been far more
surprising than before. I expose to sunlight a
sheet of cardboard saturated with tartaric acid or
with a salt of uranium, after which I roll my card-

board into a metallic tube, so that it lines the latter. I close the tube, and after a very long period of time I prove, by opening it, that the cardboard acts upon salts of silver as perfectly as it did the day it was prepared. At the ordinary temperature, this action becomes manifest only after a period of twenty-four hours; but if, after opening the tube, a few drops of water are thrown into it, and it be then closed again, and heated to forty or fifty degrees (centigrade), on re-opening the tube, and applying its orifice to a photographic paper, an impression is produced upon the latter in less than five minutes. This experiment only succeeds once, as if the photographic paper had absorbed all the light out of the tube; and to produce a second impression, the cardboard must be again exposed to light."*

This shows that *heat* and *chemical action* have an influence in these phenomena, and we know that this is very generally the case in phenomena of phosphorescence.

Mr. Draper, in his 'Human Physiology,' p. 288, describes an experiment which is closely allied to the above:—If a sheet of paper, upon which a key has been laid, be exposed for some minutes to the sunshine, and then instantaneously viewed in the dark, the key being removed, a fading spectre of

---

* This passage is not entirely in M. Nièpce's own words, but as I condensed it from his paper for the English press in 1858.

the key will be visible. Let this paper be put aside for many months where nothing can disturb it, and then in darkness be laid on a plate of hot metal, the spectre of the key will again appear.

In the cases of bodies more highly phosphorescent than paper, the spectres of many different objects which may have been laid on in succession will, on warming, emerge in their proper order.

PART II.

EMISSION OF LIGHT BY VEGETABLES.

CHAPTER I.

PHOSPHORESCENCE IN PHANEROGAMIC PLANTS.

The phenomenon of phosphorescence has, up to the present time, been very little observed in the vegetable world.

I have collected, not without much difficulty, all the observations upon this subject which appear to me worthy of confidence. Luminous plants are probably numerous, though few have been observed hitherto, and the observations we possess are somewhat scanty and uncertain.

The first discovery of a light-emitting vegetable is attributed to the daughter of Linnæus— a young damsel who was fond of setting fire on a dark summer evening, to the inflammable atmosphere which envelopes the essential-oil glands of certain *Fraxinellæ*, an experiment with which the learned François Arago was quite as delighted as the daughter of Linnæus.

A curious fact strikes us here : the first observation of vegetable phosphorescence was made by the Swedish maiden on an *orange-coloured* flower—that of the garden nasturtium (*Tropæolum majus*, fig. 6) ; and most cases of plant phos-

Fig. 6.

phorescence hitherto recorded have been observed upon flowers in which the orange and yellow tints predominate. Indeed, whether we consider phosphorescence in the mineral, the vegetable, or the animal kingdom, whether we take into consideration the colour of the substance which shines or that of the light produced, we are forcibly led to notice that of all the colours of the solar spectrum, the yellow and orange tints appear to be connected in an extraordinary manner with these phenomena.

Seated one sultry summer evening in a garden, the daughter of the illustrious naturalist observed with surprise certain *luminous radiations* emitted by the flowers of *a group of nasturtiums.* This curious observation was made more than once during twilight in the months of June and July, 1762. The girl lived long to tell her wonderful tale.*

. The same phenomenon has been witnessed by other naturalists, but almost exclusively on yellow or orange-coloured flowers. Thus, it has been seen, we are told, in the corolla of the sunflower (*Helianthus annuus*), in the garden marigolds (*Calendula*), in the two species of *Tagetes* (which the French botanists call the *Rose d'Inde* and the *Œillet d'Inde*). Phosphoric light has also been seen to be emitted from the flowers of the *Tu-*

---

* Mrs. Loudon, in her 'Ladies' Flower Garden,' p. 116, says:— " A curious discovery was made respecting this plant (*Tropæolum majus*, L.) by one of the daughters of Linnæus, who died lately at the advanced age of ninety-six. This lady, in the year 1762, observed the *T. majus*, or garden Nasturtium, to *emit sparks or flashes* in the morning before sunrise, during the months of June and July, and also during twilight in the evening, but not after total darkness came on. Similar flashes have been produced by other flowers, and it has been observed that they are always most brilliant before a thunderstorm." See also Paxton's Mag. of Botany, vol. ii. p. 195. It has been asserted that certain flowers always emit light at the periods of floration and fecundation.

G

berose, different varieties of *Nasturtium,* the *Yel-
ow Lily,* and some other plants.

The Swedish naturalist, Professor Haggern, perceived one evening a faint flash of light repeatedly
dart from a marigold (fig. 7).   Surprised at such

Fig. 7.

an uncommon appearance, he resolved to examine
it with attention, and, to be assured that it was no
deception, he placed a man near him with orders
to make a signal at the moment when he observed the light.   They both saw it constantly at
the same moment.   The light was most brilliant
upon marigolds of an orange or flame colour; but
scarcely visible on pale ones.   The flash was fre-

quently seen on the same flower two or three times in quick succession, but more commonly at intervals of several minutes; and when several flowers in the same place emitted their light together, it could be seen at a considerable distance. This phenomenon was remarked in July and August at sunset, and for half an hour, when the atmosphere was clear; but after a rainy day, or when the air was loaded with vapours, nothing of it was seen. The fact of this phenomenon only occurring when the air is dry, leads us naturally to presume that it is connected with electricity. The following flowers were observed by M. Haggern, to emit flashes more or less vivid, in this order : 1. The marigold (*Calendula officinalis*, fig. 7). 2. Monkshood, or garden nasturtium (*Tropæolum majus*). 3. The orange lily (*Lilium bulbiferum*, fig. 8). 4. The French and African marigolds (*Tagetes patula* and *T. erecta*).

To discover whether some little insects or phosphoric worms might not be the cause of this emission of light, M. Haggern carefully examined the flowers with the microscope, but no animal organisms were found. The rapidity of the flash seems to indicate that electricity has something to do with the phenomenon. The same philosopher, after having observed the flash from the orange lily, the anthers of which are a considerable distance from the petals, assured himself that the

light proceeded from the petals.   But as it is well
known that when the pistil of a flower is impreg-
nated, the pollen bursts away by the elasticity of
the anthers, and may be to a certain extent elec-

Fig. 8.

trified, M. Haggern thinks that this emission of
light by flowers is electrical, and that it is caused
by the pollen which, in flying off, is scattered
upon the petals.   Whatever we may be inclined
to think of this theory, the observations of M.
Haggern are exceedingly interesting.

The latest, and at the same time most authentic

observation of *luminous flowers* that has been made, is the following :—

On the 18th of June, 1857, about ten o'clock in the evening, M. Th. Fries, the well-known Swedish botanist, whilst walking alone in the Botanic garden of Upsal, remarked a group of poppies (*Papaver orientale*), in which *three or four flowers emitted little flashes of light.* Forewarned as he was by a knowledge that such things had been observed by others, he could not help believing he was suffering from an optical illusion. However, the flashes continued showing themselves from time to time during three-quarters of an hour. M. Fries was thus forced to believe that what he saw was real. The next day, observing the same phenomenon to re-occur at about the same hour, he conducted to the place a person entirely ignorant that such a manifestation of light had ever been witnessed in the vegetable world, and without relating anything concerning it, he brought his companion before the group of poppies. The latter observer was soon in raptures of astonishment and admiration. Many other persons were then led to the same spot, some of whom immediately remarked *that the flowers were throwing out flames.*

Some days later, on the 23rd of June, the weather having become warmer, fourteen persons again witnessed the little flashes of light on the

flowers, not only of *Papaver orientale*, but also
on those of the lily, *Lilium bulbiferum*. And before the phenomenon had ceased, upwards of a
hundred and fifty persons had been astonished
and delighted with this singular case of phosphorescence.

In the flowers observed by the daughter of
Linnæus, the phosphoric light produced was not
continuous; it manifested itself in flickerings or
sparks, which were shot out from the corolla, and
resembled somewhat those given by an electric
machine. Other observers agree with these state-
ments, and remark that the plants in question are
most luminous on calm sultry summer evenings
when the air is highly charged with electricity,
and have never been noticed to emit light when
the atmosphere is very damp.

In the phenomena remarked by Fries, the phosphorescence of the flowers always occurred between the hours of a quarter past ten and a quarter past eleven in the evening. The weather was
warm and sultry, and the luminous phenomenon
was best observed by looking at a group of
poppies without fixing the eyes upon any one
flower in particular.

But the emission of light by phanerogamic
plants is not limited to the flowers. Some naturalists assure us that the *leaves* of *Œnothera ma-
crocarpa*, an American plant, *exhibit phosphoric*

light when the air is highly charged with electricity.

The *latex* or milky juice of some vegetables becomes phosphorescent when it is rubbed upon paper, or when it is heated a little. This is extremely remarkable in *Euphorbia phosphorea,* a Brazilian species, which I believe has also been met with in Asia. If its stem be broken, and the milky juice which exudes be drawn over paper, *characters are obtained which appear luminous in the dark.* It is to M. Martins, of Montpellier, that we owe the discovery of the phosphoric property of this plant.

An emission of light has also been observed in a plant closely allied to the Palm family, and which belongs to the genus *Pandanus.* The rupture of the spatha which envelopes the flowers of this genus of plants, is accompanied by a loud cracking noise, and *a spark of light.*

The common potato in a state of decomposition sometimes emits a most vivid light, sufficient to read by. This fact was remarked some time ago by an officer on guard at Strasburg, who thought the barracks were on fire, in consequence of the light thus emitted from a cellar full of potatoes. This phosphorescence resembles that of stale fish, but it is perhaps attributable to the same cause as that of decayed wood, treated of in the next chapter.

CHAPTER II.

PHOSPHORESCENCE IN CRYPTOGAMIC PLANTS, AND EMISSION OF LIGHT FROM DECAYED WOOD.

PHOSPHORESCENCE has been rather more frequently observed in Cryptogams than in Phanerogams.

An emission of light has been observed in a pretty little plant belonging to the family of *Hepaticæ*, which grows chiefly upon schists, and resembles in miniature the royal fern (*Osmunda regalis*). From these two circumstances, the plant in question has been named *Schistostega osmundacea* (fig. 9). When this plant germinates, it gives birth to numerous confervoid filaments, which shine in a semi-obscurity with a very singular luminosity.

Unger has observed, however, that spiders' webs present very nearly the same appearance, and this circumstance has led some naturalists to believe that the shining property of *Schistotega osmundacea* may probably be nothing more than reflected light.

In the great family of *Fungi,* many cases of phosphorescence have been accurately observed, especially among the *Rhizomorphæ,*—curious vegetable organisms, resembling long thin dark-co-

Fig. 9.

loured roots (*rhizo-morpha,* in form of a root), sometimes expanding into a membraniform production, which are seen creeping between the bark and the wood of old decayed trees (willows, oaks, poplars), or shooting down into dark holes, into damp crevices, etc.

The white, flocconous extremities which constitute the *mycelium* of the species known as *R. subterranea* (fig. 10), observed not unfrequently at the

bottom and on the walls of dark damp mines or
moist caverns, on old decayed humid towers, etc.,
evolve a tranquil phosphoric light, which sometimes attracts attention by its intensity.

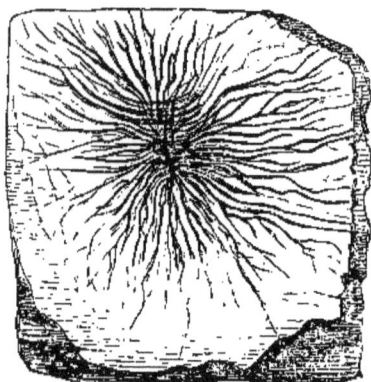

Fig. 10.

The phosphorescence of *R. subterranea* is frequent in coal mines, and has been many times
observed near Dresden. Counsellor Ehrman has
spoken with enthusiasm of its pleasing effect in
these lone and desolate places. Having once descended into one of the Swedish mines, he saw
these "vegetable glowworms" gleaming along its
walls or shining in some obscure recess.

Caverns in the granitic rocks of Bohemia are
often beautifully decked with this luminous cryp-
togam, and I am told that our English coal-mines
occasionally exhibit, by its aid, a light sufficiently
clear to admit of reading ordinary print.

But nowhere, perhaps, is the effect produced by this cryptogamic phosphorescence so exquisitely beautiful as in the mines of Hesse, in the north of Germany, where the walls of the air galleries appear illuminated with a pale light resembling that of the moonbeams stealing through narrow crevices into some gloomy recess.

Some other species of *Rhizomorpha* are supposed to be luminous, but this is doubtful. Heinzmann says he has remarked phosphorescence in *R. subterranea* and in *R. aidulæ.*

Certain experimentalists think that the light of these fungi is more brilliant in oxygen gas than in the air, and that it is extinguished in those gases which are non-respirable; whilst others, on the contrary, have asserted that though hydrogen gas, hydrochloric acid, and nitric oxide, seem to put out the light of many phosphorescent fungi, this light is not extinguished in pure nitrogen. These observations require, therefore, to be repeated with care.

Phosphorescence appears to have been first observed in large fungi at Amboine, by the botanist Rumphius, who saw light emitted from a species he has designated *Fungus igneus,* or fire-mushroom. It was afterwards seen in the Brazils by another botanist, Gardner, upon an agaric, which grows on the dead leaves of the Pindoba palm, and which has been named *Agaricus Gardneri.*

Mr. Gardner found these phosphorescent fungi in
Brazil, but it appears that a very large species
which possesses similar light-emitting properties,
is found in the Swan River colony.

A red mushroom, *Agaricus olearius* (fig. 11),

Fig. 11.

which grows at the foot of the olive-tree, in Italy,
throws out during the night a blue light, which it
spreads silently around.   This remarkable fungus
has been studied by M. Delille and M. Fabre.
According to the first-named naturalist, when
this agaric is still young, it is phosphorescent for
many successive nights, even when it is detached
from the tree, at the foot of which it is generally
to be found.   It begins to shine a little before
nightfall, continues luminous during the night,

and ceases to shine as soon as the sun rises.
The same author says that he never saw the
Agaricus olearius shine during the daytime, however dark the room in which it was kept ; and we
might remark upon this that fungi only vegetate
at night. But M. Fabre has, more recently, observed that the phosphorescence of this agaric is
not intermittent, as M. Delille supposes, and that
it shines during the day as well as by night. Exposure of the plant to sunlight appears to have no
influence whatever upon the phenomenon, " and
does not prevent its manifestation as soon as
the fungus is removed into a dark place." This
seems, however, to indicate that the sun's light
has, in reality, an influence upon the emission of
light by this fungus during the daytime, and
that the phenomenon observed by M. Fabre is
probably a case of phosphorescence after insolation—a circumstance not to be passed over slightly,
as we see further on, that a similar fact has been
observed in the insect world.

M. Fabre has also shown that the dampness or
dryness of the air does not appear to have any
influence upon the phosphorescence of *Agaricus
olearius*, unless indeed the dryness is so intense
as to desiccate the tissue of the plant. An elevation of temperature, within certain limits, does
not modify the phenomenon : below $+9°$ to $+6°$
(centigrade) the light ceases, but the phosphores-

cence recommences when the temperature is gradually raised again above this point. If the plant has been kept for some time at freezing-point, it loses its phosphoric property completely. A warmth of +48° to +50° likewise destroys this peculiar property. In other respects the emission of light by this agaric is the same under water as in the air, and pure oxygen does not appear to augment its intensity. No elevation of temperature can be observed in the parts of the fungus which shine.

The phosphoric light emitted by *Agaricus olearius* is evolved from the head (*pileus*) of this fungus: the *lamellæ* of the pileus, where the sporules or seeds are accumulated, are the seat of this extraordinary phenomenon.

The *byssoid fungi*, which penetrate the tissues of other superior kinds of fungi, or into decayed wood, are frequently seen to be phosphorescent, and the light observed is generally attributed to the decayed wood itself. This is very remarkable in old willows (*Salix*). Wood which is tender, like that of these willows, is often penetrated in all its parts by filaments of the *mycelium* of some inferior byssoid fungus, by which it acquires a peculiar fungoid smell, and becomes luminous in the dark.

This light is curious to observe under the microscope, in a dark room.

It is not exactly known to what species of fungus decayed wood owes its phosphoric light, but we can, with much probability, attribute it to the mycelium of a certain *Thelephora* which Linnæus has named *Byssus phosphorea*, placing it in the genus *Byssus*, because this illustrious naturalist was only acquainted with the filaments of the mycelium, and not with the entire plant.

Agardh, who also saw the mycelium only, classed its filaments under the name of *Mycinema*

Fig. 12.

phosphoreum, and other botanists have named them *Conferva phosphorea* and *Auricularia phosphorea*. At the present day the fungus, of which these luminous filaments constitute a part—the mycelium—is known under the specific name of *Thelephora cærulea* (fig. 12), on account of the fine blue colour observed upon the perfect plant.

It is quite possible, however, that the synonyms given above refer to more than one plant; it is very probable also that many byssoid fungi are luminous in the dark, and that this phosphoric property pertains to many other cryptogams. Adrien de Jussieu, in his ' Eléments de Botanique,' remarks that certain kinds of wood become phosphorescent when they are exposed to the damp after they have been cut in full sap. The phosphoric light emitted in this case appears to be owing to one of the byssoid fungi just named.

No *Algæ* have, if I mistake not, been described as phosphorescent, although many whilst growing under water reflect colours which perish almost immediately when the plant is removed to the air. Of this class are several species of *Cystoseira*, especially *C. ericoides*, which, though really of a greenish-olive, appears when growing under water to be clothed with the richest phosphoric greens and blues, changing momently, as the branches move to and fro in the water. Similar colours, according to Harvey, have been observed, though in a less striking degree, on some of the *Red Algæ*, and the genus *Iridæa* derives its name from this phenomenon. *Chondrus crispus* is observed to be occasionally iridescent, and at the Cape of Good Hope *Champia compressa* and *Chylocladia Capensis* present very brilliant rainbow colours, etc. " The cause of these brilliant colours," says Professor Harvey, "has not been particularly sought after."

I cannot say these pertain to phosphoric phenomena, but I have cited the above cases on account of their curiosity. They appear rather to belong to the optical phenomenon of interference.

Experiments made with a view of investigating the physical causes of the phosphorescence of decayed wood, alluded to above as owing to a minute cryptogamic organism, have consisted in placing the luminous wood into different gases, plunging it under water, etc. Bockman has proved that phosphorescent wood is as luminous in pure nitrogen and in a void, as in pure oxygen, and that its light is extinguished even in oxygen gas, if the temperature be rather high; also, that it remains luminous under water. This ingenious experimentalist has remarked that moisture exalted, to a remarkable degree, the phosphoric intensity of decayed wood, and that it appeared essential for its manifestation.

In general a certain degree of warmth and moisture, combined sometimes with a peculiar electrical state of the atmosphere—though this does not always seem essential—appear to be the most favourable conditions under which we observe vegetable phosphorescence.

H

PART III.

ANIMAL PHOSPHORESCENCE.

CHAPTER I.

EMISSION OF LIGHT BY DEAD ANIMAL MATTER.

In this section of my work I shall speak of the
emission of light by dead animal matter, before
entering upon the subject of phosphoric animals.

It is well known that when the dead bodies of
certain fish, more especially mackerels and her-
rings, are exposed to the air for a short time,
they soon become luminous in the dark. When
they are in this state, if we merely rub the finger
over the luminous surface of the dead fish, we recognize the presence of an oily substance, which
renders the finger luminous, as if it had been
rubbed upon phosphorus.

This grease, when separated from the body of
the fish by means of a knife, and placed upon a
plate of glass, continues to shine in the dark.
When examined under the microscope, not a ves-
tage of infusoria or other animalcule is seen in it,
which otherwise, as we shall see further, might
account for its luminosity.

When these dead fish are placed in sea-water, they render it luminous in the course of a few days,—the phosphorescence of the sea is however owing to a different cause,—and the water then shines in a uniform manner, *i.e.* everywhere with equal intensity: if it be passed through a filter it continues to shine as before. These facts prove clearly that this singular phosphorescence is not owing to any luminous animalcules.

Water that has been rendered luminous by dead fish loses its transparency, becomes milky, and acquires a repulsive odour; in the space of four or five days it ceases to be luminous.

Hulme, who has made numerous observations on this particular case of phosphorescence, says that the luminous greasy substance of the herring soon loses its phosphoric properties in pure water. Alcohol, acids, and alkalis also prevent its shining. Common salt and honey appear, on the contrary, to assist this phosphorescence. Sometimes also, when the latter has been extinguished by one means or another, it can be brought back again : thus, in one of Hulme's experiments, twenty-four grammes of sulphate of magnesia, dissolved in twenty-one grammes of water, and mixed with the luminous substance of the mackerel, completely extinguished its light; but if to this mixture six times its volume of water were added, it became luminous again. The same observer also remarked that the quantity of light produced in phosphorescent putrefactions diminishes as the process of putrefaction itself advances.

Cold prevents the phenomenon of phosphorescence in dead fish, but only temporarily, for the light bursts forth again with its usual intensity as soon as the temperature becomes milder. It has been also seen that this phosphorescence is accompanied by *no production of heat* in the parts which shine. We have already noticed this fact in the mineral and in the vegetable world, and we shall notice it again when speaking of luminous animals. Boiling water and high temperatures destroy the phosphorescence which occupies us here.

I have myself proved the exactness of most of the above facts whilst studying the body of a dead stockfish (*Raya*) in a luminous condition. In a short note published in the 'Comptes-Rendus' of the Paris Academy of Sciences for 1860, I have shown by direct chemical experiment that no phosphorus can be found in the luminous grease which shines upon fish. I was at first inclined to attribute their phosphorescence to the presence of some microscopic *fungi*, but at present I am more inclined to believe it is owing to some peculiar organic matter which possesses the property of shining in the dark like phosphorus itself.

The bodies of other marine animals shine after

death, none perhaps so vividly as that of the *Pholas*, a mollusk well known to those who reside on the coast. That this mollusk was luminous after death was known to Pliny, who said that it shone in the mouths of the persons who ate it; and among the moderns, Réaumur, Boccaria, Marsilius, Galeatus, and Montius have studied its phosphorescence.

Beccaria had the curiosity to ascertain how the light of putrescent fish, and that of the dead pholas, affected different colours, and for this purpose he placed in the light emitted, pieces of different coloured ribbons. The *white* came out brightest, next to that was the *yellow*, and then the *green;* the other colours could hardly be perceived. The same experiment was repeated, with similar results, on trying coloured liquids in glass tubes. We have here then another instance of the predominance of *yellow* tints over the others in cases of phosphorescence. Indeed Sir Isaac Newton, who first decomposed light into the seven rays of the spectrum, says, "The most *luminous* of the prismatic colours are the *yellow* and the *orange;* these affect the senses more strongly than all the rest together."

These experiments of Beccaria were made chiefly with the *Pholas*. A single pholas rendered seven ounces of milk so luminous that the faces of persons might be distinguished by it, and it looked as if transparent.

Beccaria and Réaumur made many attempts to render the luminosity of the pholas permanent. The best result was obtained by placing the dead mollusk in honey, by which its property of emitting light lasted more than a year; whenever it was plunged into warm water the body of the pholas gave as much light as ever.

Galeatus and Montius showed that vinegar and wine extinguished the light of the dead pholades; that a heat of 45° Réaumur (56° centigrade) extinguished this light, though it had increased in intensity up to that temperature, and that it could not afterwards be brought back again.

Many other less remarkable experiments have been noted by the above-named authors.

Most saltwater fish become phosphorescent in the dark, like those mentioned above; and concerning those which inhabit fresh water I have heard it stated that the ovaries of the carp have been seen in a phosphorescent state.

Mr. Canton has observed that several kinds of river fish could not be made to give light in the same circumstances in which sea-fish became luminous, but that a *piece of carp* made water very luminous, though the outside or scaly part of it did not shine at all.

In 1672 Boyle published a paper in the 'Philosophical Transactions,' containing observations upon shining flesh. He treats in this paper of the phosphorescence of *a neck of veal, which shone in more than twenty places, as decayed wood or putrefying fish do.*

In 1838, M. Julia de Fontenelle related in his ' Journal des Sciences Physiques et Chimiques,' a curious case of phosphorescent light observed upon the dead body of a man. Such cases of phosphorescence are not unfrequent in dissecting rooms, but often escape observation, as neither students nor professors visit these rooms at night, and when a person does happen to enter them after dark, the light he carries in his hand is too powerful to allow him to perceive the phosphoric radiations which often emanate from fragments of dead bodies lying about.

As this chapter is devoted exclusively to the phosphorescence of animal matter which has lost its vitality, I have reserved certain observations concerning evolutions of light by living subjects for a future one.

All the observations we possess regarding the nature of the light emitted by dead animal matter coincide with those of Robert Boyle, published as stated above, in the year 1672. When all the lucid parts of the shining neck of veal were surveyed at once, they made, he tells us, " a very splendid show."   By applying a printed paper to some of the more luminous spots, divers letters of the title could be distinguished.   But notwithstanding the vividness of the light, it did not yield the least degree of heat appreciable either to the touch or by the thermometer.

Boyle was often disappointed in his experiments made with a view of obtaining shining flesh at will. The luminous neck of veal was observed on the 15th of February, 1672, by one of his servants. Suspecting that the state of the atmosphere had something to do with it, he carefully noticed that the wind was south-west and blustering, the air hot for the season, the moon past its last quarter, and the mercury in the barometer at $29\frac{3}{10}$ inches.

The first distinct account that I meet with of light proceeding from putrescent flesh is that which is given by Fabricio d'Acquapendente, who says that when three Roman youths residing at Padua had bought a lamb and had eaten part of it on Easter-day, 1592, several pieces of the remainder which they kept till the day following shone like so many candles when viewed in the dark. Part of this luminous flesh was sent to Fabricio d'Acquapendente, who was then Professor of Anatomy in Padua. He observed that both the lean and the fat shone with a white kind of light, and that some pieces of kid's flesh which had lain in contact with it were luminous, as well as the fingers of the persons who touched it. Those parts shone most which were soft to the touch, and which appeared more or less translucid when held before a lighted candle.

The next account of a similar appearance was described by Bartholin, the celebrated Danish philosopher, as seen at Montpelier in 1641. A poor woman had bought a piece of flesh in the market, intending to make use of it the day following; but happening not to sleep well that night, and her bed and pantry being in the same room, she observed so much light come from the flesh as to illuminate all the place where it hung. This flesh was shown to many persons as a curiosity, and kept *till it began to putrefy, when the light vanished.*

After these come Boyle's observations alluded to above.

It has often occurred to me that this singular production of light in dead animal matter *precedes putrefaction;* no disagreeable smell is observed until the luminous appearance has lasted some time. Boyle was also aware of this, for he says, " Notwithstanding the great number of lucid parts," referring to his neck of veal, *" not the least degree of stench was perceivable to infer any putrefaction."*

Water does not destroy the phosphorescence of dead animal matter; but alcohol, acids, etc., soon extinguished it. According to Boyle's experiments, a piece of shining flesh shone less, but did not lose its light, when placed in the exhausted receiver of an air-pump; "but," he adds, "by the hasty increase of light that disclosed itself in the veal upon admitting the air into the exhausted receiver, it appeared that the decrement, though slowly made, had been considerable." The luminosity of flesh generally lasts about four days, after which putrefaction sets in rapidly.

A peculiar mucilaginous substance, or mucus, is sometimes seen about spring, on the damp ground near rivulets or stagnant pools in the fields, which, from the circumstance of its being occasionally phosphorescent at night, has been regarded, since the middle ages, as having some connection with shooting stars. The Belgian peasants call it " the substance of shooting stars."

I have sketched the history of this curious substance in the 'Journal de Médecine et de Pharmacologie' of Bruxelles, for 1855. It was analysed chemically by Mulder, and anatomically by Carus, and from their observations appears to be the peculiar mucus which envelops the eggs of the frog. It swells to an enormous volume when it has free access to water. As seen upon the damp ground in spring, it was often mistaken for some species of fungus; it is however simply the spawn of frogs, which has been swallowed by some large crows or other birds, and afterwards vomited, from its peculiar property of swelling to an immense size in their bodies. From the fact of this mucilaginous matter having been sometimes observed in a luminous state at night, the Dutch and Belgian peasants imagine that it has been dropped upon the ground by some passing shooting star; and in Mulder's account of its chemical composition, given by Berzelius in his 'Rapport Annuel' (French edition), he distinguishes it by the designation of "mucilage atmosphérique."

My attention was called lately to a case of luminous urine, observed by a friend of mine, in 1859, during a warm summer in Paris. It was observed to shine with a slight phosphoric light when stirred. I was at first inclined to attribute this to a phenomenon of reflection, but I find that the same fact was observed many years ago by two well-known medical men, Reiselius and Pettenkofer, one of whom witnessed this phosphorescence in November, and the other in March. It appears, therefore, evident that urine may become luminous in certain circumstances with which we are not acquainted. The old chemist Lemery has, moreover, remarked that urine is sometimes phosphorescent.

CHAPTER II.

EMISSION OF LIGHT BY INFERIOR ORGANISMS.
PHOSPHORESCENCE OF THE SEA.

I SHALL now enter upon the subject of Phosphoric
Animals; *i.e.* of the phenomenon of *phosphorescence
in living animal organisms*. And in the first place
I shall draw attention to a curious fact. With the
exception of a few more or less doubtful cases, to
which I shall allude at the end of this part of my
work, the faculty of producing light seems, in the
animal world, to cease with the class of insects.
But, on the other hand, from insects downwards,
there is scarcely a section of the animal world but
which furnishes us with some self-luminous beings.
Thus decided cases of phosphorescence have been
and are frequently observed, in *Infusoria, Rhizopodes, Polypes, Echinoderms, Annelides, Medusæ,
Tunicata, Mollusks, Crustaceans, Myriapodes*, and
Insects.

It would indeed require volumes to describe
each luminous animal belonging to these numerous tribes. I shall not attempt it here, but I

shall call attention to the most remarkable of them. Their zoological descriptions, *i.e.* their nature and habits, can be found in other works.

A vast number of these inferior organisms render the waters of the ocean luminous in every latitude. The little beings classed in the genus *Noctiluca*, which resemble the larger kinds of Infusoria, but belong in reality to the class of *Rhizopodes*, play an important part in the illumination of the sea. *Polypes*, *Medusæ*, a whole host of *Infusoria*, some *Worms*, and some small *Crustaceans*, contribute also to the beauty of this phenomenon.

In the years 1749 and 1750, Vianelli and Grixellini, two Venetian naturalists, discovered in the waters of the Adriatic, considerable quantities of an animalcule evidently possessed of luminous properties. They immediately attributed to this singular being, the cause of the phosphorescence of the sea, a phenomenon which to that day had remained a mystery. This animalcule received from Linnæus the name of *Nereis noctiluca*.

In 1776, Spallanzani was made aware of the self-luminous properties of a Mediterranean blubber, *Pellagia phosphorea*, and at the commencement of the present century Viviani made known the following fifteen species of phosphoric animals, *Asterias noctiluca*, *Cyclops exiliens*, *Gammarus caudisetus*, *G. longicornis*, *G. truncatus*, *G. heteroclitus*, *G. crassimanus*, *Nereis mucronata*, *N. radi-*

ata, Lumbricus hirticauda, L. simplicissimus, Pla-
naria retusa, Brachiurus quadruplex, and *Spirogra-*
phis Spallanzanii. Some of these were found off the
coast of Genoa in 1805. Scoresby and Riville soon
recognized other phosphorescent species, which
they dredged from the ocean in their voyages.
Macartney made known, in 1810, the luminous
Medusa scintillans, M. lucida, and another curious
little being closely allied to Medusæ, called *Beroe
fulgens.* Peron and Lesueur, in their voyage from
Europe to the Isle of France, discovered the *Pyrosoma Atlantica* (fig. 12 : 1, the entire animal mag-

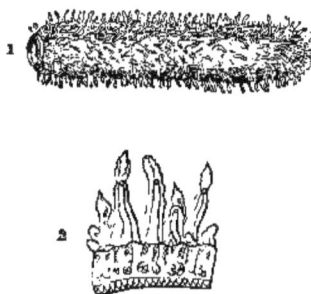

Fig. 12.

nified; 2, the phosphorescent surface of the body,
magnified 300 diameters), one of the most curious
of animals. It belongs to the tribe of *Tunicata ;*
each individual resembles a minute cylinder of
glowing phosphorus ; sometimes they are seen
adhering together in such prodigious numbers,
that the ocean appears as if covered with an enor-

I

mous layer of shining phosphorus or molten lava.
These singular productions of nature are met with
between 19° and 20° of longitude east of Paris,
and 3° and 4° north latitude.

The no less curious animals belonging to the
genus *Salpa* (fig. 13, *Salpa cristata*: 1, an isolated individual; 2, five *Salpæ* united as they
swim), also classed in the *Tunicata*, abound in the

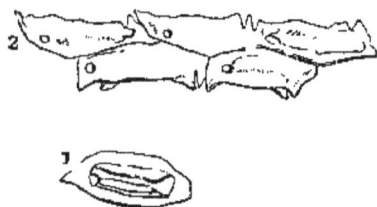

Fig. 13.

Mediterranean and warmer parts of the ocean.
They are often phosphorescent. They also swim,
adhering together in vast numbers; their phosphorescence resembles the light of the moon on
the still waters, and they give rise to what is termed
by the French a *mer de lait*, or sea of milk.

Sir Joseph Banks, in his voyage from Madeira
to Rio Janeiro, discovered the little crab *Cancer
fulgens*, a species said to be very phosphorescent.
In nearly the same latitude that this discovery
was made, *Medusa pellucens* was met with; its
phosphorescence is described as resembling a
flash of lightning.

In 1810, M. Suriray showed that the phosphorescence of the sea in the English Channel was owing exclusively to *Noctiluca miliaris*, a very minute Rhizopode, which has since been studied by

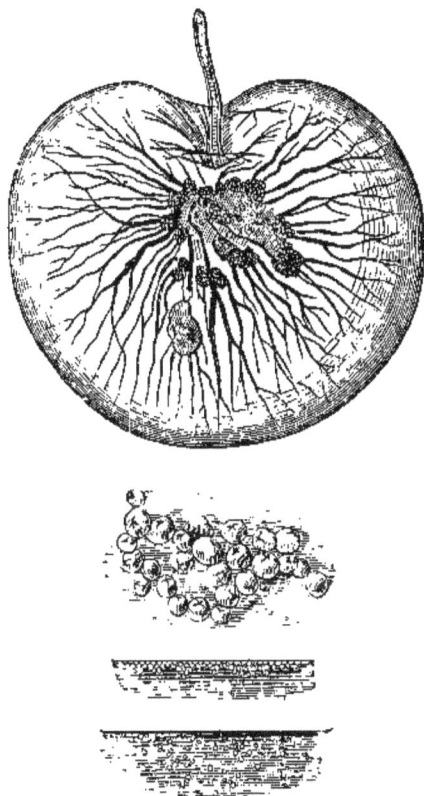

Fig. 14.

M. de Quatrefages, of Paris, and Dr. Verhaghe, of Ostend. (Fig. 14, *Noctiluca miliaris*: 1, The animal highly magnified; 2, less magnified; 3, as seen with a pocket lens on the surface of a glass of water; 4, as they descend through the water, glowing with phosphorescent light, when the glass is shaken.)

In the year 1830, Michaelis, a distinguished professor at Kiel, was, according to Humboldt, the first to make known the existence of *Phosphoric Infusoria*. He first observed the phenomenon of phosphorescence in a species of the genus *Peridinium* (fig. 15), a ciliated animalcule, and

Fig. 15.

afterwards in *Prorocentrum micans*,* and in the

* It is exceedingly probable that this animalcule will be placed among the Rhizopodes; and the same remark may apply to many now-called *Infusoria*. In this microscopic class of animals, as it undergoes fresh investigations, the species are continually being removed and placed in higher genera, families or classes. Thus the *Rotifera* are now classed among the Annelides.

Rotifer which he has named *Synchata Baltica*, indicating that it is found in the Baltic Sea. The naturalist Focke has since found this remarkable little being in the lagune of Venice.

Ehrenberg has described the following species of self-luminous Infusoria existing in the Baltic : —*Prorocentrum micans*, an oval-shaped animalcule with a long cilium, and containing many transparent nucleoles (fig. 16) ; *Peridinium Michaelis*,

Fig. 16.

P. micans, P. fusus, P. furca, P. acuminatum, Synchata Baltica, and a species of *Stentor* (*S. igneus?*).

Peridinium Michaelis (fig. 17), named after the distinguished naturalist already alluded to, is not unlike a very minute Florence flask filled with plums, standing upon two legs, and having a ciliated belt round its middle. It is perfectly invisible to the naked eye. The other species are very odd-looking creatures, which it is impossible to describe here.

Fig. 17.

The interesting *Synchata Baltica* (fig. 18) belongs to the tribe of *Hydatinea,* of which many individuals are common in our ditches and stagnant freshwaters.

The largest of these phosphorescent Infusoria described by Ehrenberg, are about one-eighth or a line, the smallest are from a forty-eighth to a ninety-sixth of a line in size. They offer a

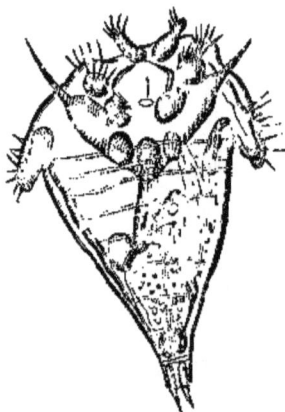

Fig. 18.

magnificent spectacle when placed under the microscope in a dark room.

Ehrenberg has likewise studied certain species of *Photocharis,* marine animalcules which resemble *Nereis* among the worms; and, when seen under the microscope, appear like minute strings of lighted sulphur. The phosphoric light they emit is of a greenish-yellow, similar to that of the common glowworm. *Oceana hemisphærica,* another minute creature described by the same naturalist, appears " enveloped with a shining crown."

At the present day it appears doubtless that certain small crustaceans are occasionally phosphorescent; especially the species *Cancer fulgens* already mentioned, and *Cyclops quadricornis.*

Many mollusks have also been seen to shine at night, especially certain *Pholades* which bore into ships, etc., and, according to some authors, certain terrestrial species. Concerning these mollusks, it is well known that they shine intensely after death; but do they shine when alive? I have not been able to solve this query. The other animals alluded to have been frequently seen, and many of them by myself, in a shining condition during life.

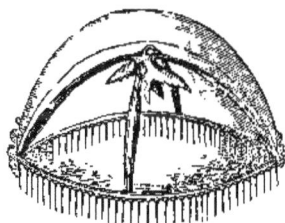

Fig. 19.

The different species of *Cephalopoda* which live near the shore, and some *Pteropoda,* have been seen to be self-luminous. The same may be said of the curious organisms called *Biphores, Dyphises, Phy-*

salia, Salpa, certain *Nereids;* and among the Star-fish (*Asterias*), I should mention the *Ophiura,* or Sand-stars.

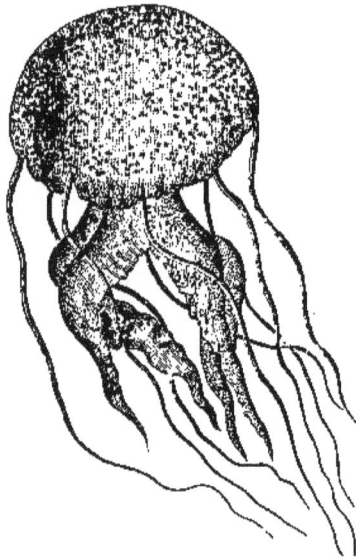

Fig. 20.

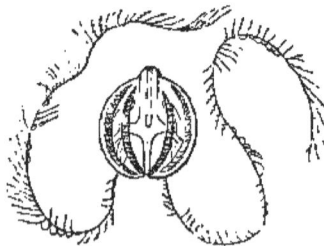

Fig. 21

Medusæ (figs. 19, 20, 21, and 22) and *Cyaneæ* possess this luminous property to a very high degree, and so do many polypes, such as the *Penna-tula*, known to naturalists as the *Phosphorescent Sea-pen*, and in French as *Plumes de mer*. Some

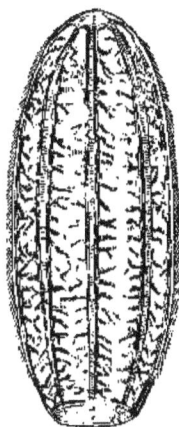

Fig. 22.

species of *Veretilla* and *Virgularia* (Sea Rush) are known to be self-luminous. The thick branchy

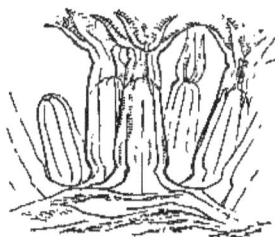

Fig. 23.

Alcyonidium gelatinosum (fig. 23), which is only found attached to rocks situated in deep water,

has been remarked to emit phosphoric light at certain seasons of the year.*

Among the Rhizopodes, the species named by
Ehrenberg *Mammaria scintillans*, now better
known as *Noctiluca miliaris*, which is never larger
than the head of a small pin according to Humboldt (I never saw one half so large), " offers to
the microscopical investigator of Nature, the magnificent spectacle of a starry firmament reflected
in the sea." Humboldt tells us, in one of his
works, that his body has remained luminous for
an hour together, covered as it was by *Noctiluca*,
after bathing in the waters of the Pacific.

Noctiluca miliaris is very common in the English Channel; and I have found this species in such
prodigious numbers in the damp sand at Ostend,
that on raising a handful of it, it appeared like so
much molten lava.

In the year 1854, the number of marine animals
known to be endowed with phosphorescent properties during life, amounted to upwards of a
hundred distinct species, all invertebrata.

MM. Edoux and Soulezet, two French natu-

* It would certainly be interesting to introduce some of these
luminous animals into the marine aquariums which are much
in vogue at present. For my own part, I have often been delighted with the phosphorescent spectacle some of them present in
a small flask. The species figured in the text are perhaps among
the more worthy of notice in this respect, viz. fig. 19, *Thaumantius
pilosella;* fig. 21, *Cydippe pileus;* fig. 22, *Beroë Forskallii*, etc.

ralists, who made a scientific voyage round the
world in the ship 'La Bonite,' observed that certain small phosphorescent Crustacea sometimes
secrete a peculiar phosphorescent matter, and
that when they are irritated, they send forth magnificent flashes of light. Other crustaceans did
not appear to possess this faculty. These two
gentlemen collected a certain quantity of the
phosphoric substance; they found it to be "*yellowish*, viscous, and soluble in water," communi-
cating its luminous property to this liquid, but
only for an instant or two. It lost its luminosity
when it had been separated for a few moments from
the body of the animal.

These naturalists have also observed the phenomenon of phosphorescence in certain Pteropoda
and Cephalopoda; they believe that these animals
possess a peculiar phosphorescent substance like
that of the Crustacea, which, from their observations, appears to be continuously and uniformly
luminous as long as the animals live, but ceases
to be so when they die.

When the phosphorescent Infusoria, of which
I have spoken, are exhausted and cease to emit
any light, their luminous faculty can be restored
for a while by exciting them with a drop of dilute
acid or alcohol mixed with the sea-water; but
this experiment soon kills them.

Humboldt affirms that having placed certain

Medusæ upon a tin plate, he observed that whenever he struck this plate with another metal, the slightest vibrations of the tin were sufficient to render the animal completely luminous many consecutive times. It has been observed also that these Medusæ emit a more intense phosphorescence when they are placed in a galvanic circuit; but the electric current must not be too powerful.

The experiments undertaken by M. Suriray a Havre, by Professor Ehrenberg at Heligoland, by M. de Quatrefages at Boulogne, and by Dr. Verhaghe at Ostend, have added considerably to our knowledge of the emission of light by *Noctiluca*.

All mechanical or chemical agents that bring about a *contraction of tissue* in these animalcules excite their phosphoric quality. A drop of weak acid or alcohol, a shock given to the glass which contains them, immediately renders every individual luminous. If a few teaspoonfuls of *Noctiluca* be collected upon a filter, the light they emit is powerful enough to enable us to read at a distance of nine inches and a half. When the bulb of a small and very sensitive thermometer is plunged into this little heap of *Noctiluca*, it is found, that although these small beings are in full life, not the slightest elevation of temperature can be observed during the emission of their light.

A curious observation has been made by Professor Ehrenberg. By submitting his " Mammaria" (*Noctiluca*) to a magnifying power of thirty diameters, he saw some of them become brilliant on one part of their body only, others on many points, and others again on their whole surface. On increasing the magnifying power from sixty to one hundred and forty diameters, more and more brilliant points became luminous. The light seemed concentrated in them, whilst the homogeneous luminosity of the animalcule disappeared. Ehrenberg considers these brilliant points to be so many light-emitting organs.

CHAPTER III.

PHOSPHORESCENCE OF THE EARTHWORM.

In the year 1840 M. Forester wrote to the Academy of Sciences at Paris, stating that during a dark rainy night he had seen a great number of *Lumbrics,* or earthworms, shining with a white phosphoric light, which he compared to that of iron heated to a white heat.

When this letter was communicated to the Academy, the distinguished naturalist M. Audouin rose and said, that to his knowledge no authentic case of phosphorescence in earthworms had ever been made known, but that he could cite numerous cases where these worms had been mistaken for *Scolopendra,* some species of which are known to be phosphorescent.

Upon this occasion M. Duméril, lately one of the greatest ornaments of the present Institute of France, said that he happened to know of two very authentic observations, emanating from two eminent naturalists, concerning the emission of light by earthworms, properly so called (*Lumbricus*); the first of these was made by Flaugergues, who had observed the phenomenon of phosphorescence in *Lumbrics* for some years consecutively, and always in the month of October; namely, in 1771, 1775, and 1776. Flaugergues had, moreover, remarked that the light was emitted principally from that portion of the body in which are situated the external organs of reproduction. The second observation is owed to the naturalist Bruguière; his note, which was inserted in the ' Journal d'Histoire Naturelle' (vol. ii. p.267), is entitled " Sur la Qualité Phosphorique du Ver de Terre en certaines circonstances."

Since then, M. Audouin himself has been convinced of the fact by some curious observations made known to him by Professor Moquin-Tandon, one of the present members of the Academy of Sciences. These observations are well worth recording.

The last-named *savant*, together with M. Saigey, remarked one warm summer evening in the year 1837, a number of small phosphorescent animals in a garden-walk at Toulouse. Both M. Saigey and M. Moquin-Tandon ascertained positively that these animals belonged to the genus *Lumbricus*. They were from forty to fifty millimètres long. The light with which they shone was nearly white, and resembled that of a bar of iron heated to a white heat. When one of these worms was trodden upon and crushed, the phosphorescence spread out upon the ground, producing a long train of light, as if the earth in this place had been streaked with a piece of phosphorus.

Each of these *Lumbrics* was remarked by their observers to have a well-developed *clitellum*, which proves that the worms under inspection were adults and that it was their period for coupling. M. Moquin-Tandon preserved some of these worms for many days, and observed that their luminous property resided in the sexual swelling or *clitellum*, and that their phosphorescence ceased immediately after copulation.

The editor of a French periodical,* in which my *brochure* on Phosphorescence was reprinted without my consent soon after it appeared in France, received shortly afterwards a letter, dated 18th November, 1858, and signed M. Adrien, of Pont Saint-Esprit, in which the writer declares that, having read my papers, the paragraphs which treat of the phosphorescence of *Lumbrics* had reminded him of an observation he had made about three years ago.

" One summer's night after a rainy day," says the writer, "I saw the ground sparkling with a whitish phosphoric light whilst sprinkled with warm urine, and I recognized at the same time

---

* 'L'Ami des Sciences.' Paris, 1858.

the presence of numerous small worms. . . . The phenomenon was so curious that I took up some of these worms and carried them into the house to examine them by the light of my lamp. I immediately recognized them to be small *Lumbrics,* about fifteen millimètres long. Returning again into my garden, with a lantern, I saw at the same place many *Lumbrics* crawling upon the ground with their usual slow and regular mode of progression. But they showed no light; and when the lantern was put out, their presence could not be recognized. But as soon as they were in their turn sprinkled with warm urine, the phosphorescence of their entire bodies shone forth and illuminated their wriggling movements."

The writer of this letter says he has since repeated the experiment many times, and he asks if the phenomenon ought to be attributed to the saline matter contained in the urine, or if the warmth of this liquid is alone necessary to occasion the phosphorescence of *Lumbrics.* This might have been easily ascertained by using pure warm water in the experiment; but the author of the letter apparently did not think of it. His observation, as we have given it in his own words, tends to prove that violent muscular contraction excites an increase of phosphorescence in the earthworm as in the *Noctiluca* mentioned among animals in the preceding chapter; and the fact

K

concerning the earthworm appears to have been observed by M. Vallot, of the Academy of Dijon, as early as 1828. I may add here, that I distinctly remember witnessing, when quite a child, the phosphorescence of the earthworm; the light appeared connected with the slimy matter that covers the animal's body. It was whilst digging at night, in a large dunghill, for worms to supply baits for a fishing excursion that my schoolfellows and myself turned up many hundred *Lumbrics* in a highly luminous condition ; but I cannot recollect in what month this happened.

CHAPTER IV.

PHOSPHORESCENCE OF SCOLOPENDRA.

It is well known that the centipedes belonging to the genus *Scolopendra*, and the class of *Myriapoda*, present us with at least two self-luminous species.

In common with the earthworm, *Scolopendra* emit phosphoric light of a greater intensity at the time when the functions of reproduction are about to be performed.

We must register here another anecdote :—

On the 16th of August, 1814, about nine o'clock in the evening, some persons came to M. Audouin at Choissy-le-Roi, near Paris, where he was passing his vacations, and called his attention to a curious fact. They had seen, they said, an immense number of "earthworms" in a chicory-field not far distant, and these "worms" shone with a light that could only be compared to that of a piece of coal white-hot. One of these was brought in a flowerpot to M. Audouin. It was evidently a *Lumbric* ; but, at the same time, this *Lumbric*

was not phosphorescent. All present were greatly
surprised, and so was M. Audouin, when the
latter, removing some earth from the flowerpot,
soon discovered six small *Scolopendra* belonging
to the species *S. electrica* of Linnæus. Their
phosphoric light was indeed vivid enough.

Going afterwards into the chicory-field, M.
Audouin observed this phosphorescence on a
grand scale. At first he saw only a few streaks
of light upon the soil; but, having ordered some
of the earth to be dug up, the spectacle that presented itself was truly magnificent. The up-
heaved soil appeared everywhere sprinkled with
phosphoric radiations, and if some of it was trodden upon or rubbed between the hands, streaks
of light were produced which remained visible for
eight, ten, and twenty seconds.

Many persons have witnessed the luminous
phenomenon of *S. electrica,* and their observations
coincide precisely with those just related.

Scolopendra electrica (fig. 24) and *S. phosphorea*

Fig. 24.

are the only two species that are known with certainty to be highly phosphorescent. But it is

probable that future observations will furnish us with others. The *S. electrica* of Linnæus is not uncommon in England, Belgium, France, etc. Its light is seldom seen, in consequence of its habit of living in holes in the soil; but it is sometimes to be met with in outhouses or crawling along some secluded pathway, leaving a track of phosphoric matter behind. It is about an inch and a half long, its diameter being scarcely more than one-tenth of an inch; its colour is a dusky brown, and its legs, which are one hundred and forty in number (seventy upon each side of the animal's body), are of a yellowish hue.*

We know very little of *S. phosphorea*, which appears to be a native of Asia.

Some authors state that *S. electrica* is only phosphorescent when in motion, and that its light cannot be discerned when the creature is at rest; it is particularly brilliant, however, when the animal is disturbed or irritated.†

Macartney has made some extremely curious observations on the phosphoric properties of our English *Scolopendra*. It results from his researches that the *S. electrica* is capable of secreting a luminous fluid (like the little Crustacea observed by Edoux and Soulezet, as I have already

---

* I have already alluded to the *yellow* colour as being apparently so intimately connected with phosphoric phenomena.

† We have seen that this is the case with *Noctiluca.*

noticed), and that this fluid can be communicated by the centipede to every part of its tegument. He has remarked also—and his observation has since been confirmed by Kirby and Spence—that this fluid can be received upon the hand, where it will remain luminous for some seconds.   But the most curious of Macartney's observations is this— he believes that this peculiar luminous substance of *S. electrica* does not shine in the dark unless it has been previously exposed to the solar rays.

This is certainly a remarkable observation, and if it should be confirmed by future investigations, it will constitute a very important feature in the phenomena of animal phosphorescence.   We shall see presently that a similar observation has been made with regard to the phosphorescent substance of a luminous insect belonging to the genus *Lampyris.*

CHAPTER V.

PHOSPHORIC INSECTS.

PHOSPHORESCENCE has been observed with certainty in a considerable number of insects belonging to the numerous family of *Coleoptera*, and in some belonging to the family of *Hemiptera*. We possess also some doubtful observations of this kind regarding certain *Lepidoptera* and *Orthoptera*.

First among luminous *Coleoptera* I must mention the genus *Lampyris*, to which belong our own *Glow-worms* (*Vers-luisants* or *Lampyres* of the French).

There are many species of *Lampyris*. No insects, perhaps, have given rise to more poetical sentiments among English authors, some of whom have termed them " stars of the earth," " diamonds of the night," etc.: pseudonyms they owe to their faculty of emitting a tranquil phosphoric light, by which they illuminate and decorate our hedgebanks on fine summer nights. For, if we

examine them in the daytime, these insects, as
every one knows, are not characterized by any
extraordinary feature, nor do they astonish us by
their beauty.

Lampyris noctiluca (fig. 25) is the species most
abundant in England, Belgium, Germany, and the

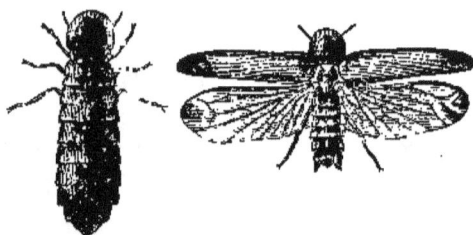

Fig. 25.

north of France. We all know it well. We have
all admired it silently shining on the fresh green
sward of the country, and we all value this insect
for the agreeable *souvenirs* which it calls forth as
we contemplate its soft light. These little shining
beings remind us of our younger days. They
were shown to us in our early childhood, and we
have been taught to look upon them as something
mysterious. " Those sparks in the grass, what are
they? Insects! But the light!" How often
have we not heard such questions. Or this, again :
" Tell me, then,—you, a philosopher,—what is it,
definitely, that produces the light of the glowworm?" To which we can only reply, " Those
sparks in the grass have excited the inquisitive-

ness of philosophers, no less than the curiosity of children ; but that which is a mystery to the latter is still a secret, or nearly so, for the others."

Lampyris hemiptera is a rarer insect with us, but it is met with, nevertheless, now and then. It is black and small, its body being rather elongated, its elytra or wing-cases conical, and the extremity of its abdomen yellow.

Lampyris italica, both sexes of which are winged,—which is not the case with the two foregoing species, the females of which have no wings, —sheds a very brilliant light ; it inhabits Italy and Southern Europe, but has been accidentally met with in England and Belgium. *Lampyris splendidula* and *L. mauritanica* are found in the South of France, and *L. corusca* in Russia.

An error, that has become popular, has held ground regarding glow-worms. It has been stated and persisted in, that the *males* of the different species of *Lampyris* have not the faculty of emitting light. Now it has been shown long ago that this opinion is inexact. An English naturalist, Ray, was the first to observe that the male of our *L. noctiluca* shone in the dark. Geoffroy afterwards found that this male insect has four small *luminous* points, two upon each side of the abdomen : and Müller confirmed his observation. The male insect of *L. splendidula* and that of *L. hemiptera* show a very brilliant light when

thcy fly. However, both the males and females
of glow-worms possess the faculty of extinguishing or emitting their light, seemingly at will.

The light of tho females is emitted from the
last three segments of the abdomen. On the last
segment of all we find, in *L. noctiluca,* two small
luminous points, more brilliant than the rest of
the segment. The light of *L. italica* is very
bright. Both sexes fly, and these insects are not
uncommon in Italy. Whilst flying through the
evening air they produce a very pretty effect; at
first sight, the stars appear to be moving about in
all directions. We are told that it was formerly
a custom among the Italian youths to decorate
the hair of their mistresses with these " diamonds
of the night," which were probably less expensive
than pearl necklaces, and evidently surpassed the
brightness of the mineral diamond.

We know that the light of glowworms is transmitted to us directly from the body of the insect,
for it does not possess the properties of reflected
light. It has been remarked, that the light of *L.
noctiluca* is refracted like that of the sun or the
stars, when it passes from one medium to another
of greater or lesser density. But it has never
been analysed prismatically, so that we cannot say,
at present, whether it is possessed of any peculiar
properties.*

* In 1808, Wollaston discovered what are called the *dark lines*

Matteucci has made a number of experiments upon *Lampyris italica,* with a view of proving that the phosphorescence of the glowworm is a phenomenon of combustion ; and, however erroneous this opinion may hereafter appear, M. Roberts had already professed it as the result of his own experiments a year before M. Matteucci. This part of my work is not devoted to the dis-

of the solar spectrum ; twelve years afterwards Fraunhofer determined the position of these lines, which are simply breaks or interruptions in the coloured spectrum, with great accuracy. Whenever light comes from *the Sun,* as solar light, as planetary reflected light, as lunar light, or as light reflected from the clouds, the number and position of these lines are the same. But when light comes from other sources than the sun—from other suns, for instance, from the star *Sirius*—these dark lines of the spectrum differ. We find that the dark lines in the spectrum of *Sirius* differ from those of *Castor* or from those of our *Sun.* This difference, which was first indicated by Fraunhofer, was afterwards confirmed by Professor Amici, who also showed that in fixed stars that have an equal and perfectly white light, the dark lines are not the same. We also know that the *specific character* of the source of light, *i.e.* its nature, has an influence upon these lines of the spectrum. Thus the light of the *electric spark* and that of *incandescent solid bodies* exhibit great diversity in the number and position of Wollaston's dark lines. However interesting the inquiry may appear, this knowledge has not yet been applied to the light emanating from *luminous animals,* nor indeed to any phosphorescent substance in the strict sense of the term.

(Since the above was written, Kirchhoff, of Heidelberg, has discovered that the dark lines of the solar spectrum are owing to the presence of metals in the sun's atmosphere.)

cussion of theoretical points; I must, however, expose a few facts.

Matteucci himself observes that glow-worms possess a substance which gives forth a brilliant light *without any sensible heat;* and that this light is visible after the animal has been torn to pieces; that it persists for some time after death.

Those animals which furnish us with examples of rapid and energetic combustion, such as birds, by their respiration, possess also a high degree of animal heat, when compared with animals in which respiration is less energetic. With reptiles, where combustion, due to respiration, is comparatively incomplete, we find on the contrary, an animal heat of a low degree, dependent upon the temperature of the place they inhabit. In accordance with this observation, if the light of *Noctiluca, Pyrosoma,* or *Lampyridæ* was owing to combustion, their animal heat would be considerably higher. Experiment, however, shows us the contrary.

According to Matteucci, also, the phosphorescent substance, when extracted from the insect and placed in hydrogen gas or carbonic acid, "ceases to shine in thirty or forty minutes." *Thirty or forty minutes* is a very long time indeed, for a *combustion* to continue in gases that are devoid of the faculty of supporting it. As this experiment was very carefully verified by the author, we can assert without danger, that in gases which extinguish combustion the luminous substance extracted from glowworms shines for thirty or forty minutes.

The same author next remarks, that in oxygen gas, the phosphorescence of the luminous substance shines three times longer. Only *three times*. But this may be due to a difference of vitality in the different individual insects submitted to examination. Heat increases the light of glowworms ; too great a temperature destroys it.

On the whole, the experiments of M. Mattcucci, made with remarkable delicacy, can lead to no conclusion which tends to establish the nature of the phosphorescence of the *Lampyridæ ;* unless, indeed, this conclusion be that the phenomenon in question is not directly owing to combustion.

With the exception of a single one, the same may be said of those undertaken by M. Roberts, and communicated to the ' Annales des Sciences Naturelles' in December, 1842. M. Roberts has, however, noted a fact similar to the observation of Mr. Macartney, regarding the luminous substance of the *Scolopendra*. I shall give it in his own words :—

"If a female *Lampyris* be divided into two transversal halves, the light spread around by the abdominal portion disappears in about half an hour. But by placing this same portion near a

candle, the light reappears, and, strange to say, is
not extinguished in thirty-six hours." He adds
to this, " It is in vain that I endeavoured to make
it appear again by the same process ; this singular
phenomenon seems only to take place once."

This observation seems to be a complete representation of phosphorescence after insolation, such
as we observe in mineral substances. It would
certainly be interesting to ascertain whether solar
light has any influence upon the phosphorescence
of glowworms. We must not forget, however,
that the luminous substance of the glowworm is
capable of shining for some time after death. If
M. Roberts was not aware of this, his observation
cannot have much weight.

I must register here an interesting experiment
made some years ago by Dr. Lallemand :—One
fine summer evening, M. Bérard, of Montpellier,
had invited to his house a number of professors
and naturalists. Dr. Lallemand, who was present,
caused them to witness a very curious phenomenon. He placed upon his hand a female glowworm (*Lamypris noctiluca*), and stretched his arm
out of the drawing-room window, which opened
into the garden. Very few instants elapsed before a male *Lampyris* flew into the doctor's hand
and immediately coupled with the vermiform female which he held. But as soon as the act was
accomplished, the light of the female was extinguished completely.

This experiment was witnessed by many eminent *savants,* among whom were MM. Bérard, Dujès, Dubreuil, Balard, and Moquin-Tandon.

M. Schnetzler, of Vevey (Switzerland), made some experiments upon *Lampyris noctiluca* in 1854. He attributes the light of these insects to the combustion of phosphorus, which he thinks he has found in the greasy tissue which constitutes the luminous organ of the insect.* Phosphorus probably exists in the luminous tissue, but only in the state of *phosphates.* By heating this tissue with nitric acid until complete dissolution was obtained with destruction of the organic matter, M. Schnetzler procured a solution showing those chemical reactions which characterize *phosphates.* But that does not prove that *phosphorus* in a free state exists in this tissue.

An English chemist, Mr. Thornton Herapath, asserts, on the other hand, that the most delicate analysis did not show the slightest quantity of phosphorus (as *phosphate*) in the bodies of those insects.

Here, then, we have one observer inferring the existence of free phosphorus after finding phosphates, and another denying the existence of free phosphorus after seeking for phosphates only. Mr. Herapath thinks, in his turn, that the light of the glowworm is due to a compound of hydro-

* 'Archives des Sciences Physiques de Genève,' Nov. 1855.

gen and carbon, secreted by a particular gland,
organized for that purpose. I believe this car-
buret of hydrogen exists in the tissue of the
Lampyridæ, where it forms, with other substances,
that greasy matter that all observers have remarked in the luminous tissue, not only of *Lampyridæ*, but of other phosphorescent insects. But I
doubt that the phosphorescence of these insects
is owed to combustion. However, I have given
this opinion its due in the theoretical part of my
work.

M. Schnetzler* has brought forward some other
observations on *Lampyris noctiluca*, which appear
worthy of note. It is generally believed that the
light of the glowworm is not visible in the daytime, for the simple reason, perhaps, that a light
sixty times stronger than another prevents our
perceiving the latter. But, according to the author just named, if the inferior posterior portion
of the abdomen of a female glowworm be opened,
we perceive a yellowish-white substance which
emits a very feeble light *during the day*.

Although the light of the glowworm appears
to be in direct submission to the will, or rather to
the *instinct* of the insect during its life, and can
therefore be extinguished more or less at certain
intervals, it is not less true that this light persists
for some time after death, and even after the lumi-

* *Loc. cit.*

nous substance has been extracted from the insect's body. According to Carus,* the light, when extinguished in the dry luminous matter, reappears when the latter is damped with water. The naturalist Boitard has made a similar assertion; speaking of the glowworm's light, he says: " Il paraît qu'ils peuvent à volonté augmenter ou diminuer cette singulière lumière, qui disparaît lorsqu'ils sont morts, *mais seulement par le dessèchement.*"†

We are thus led to infer that the luminous substance of glowworms is permanently phosphorescent even after death, if the tissue of the luminous organ be kept dry, and not decomposed by chemical means.

When the insect, dead or living, is plunged into boiling water, its light is extinguished suddenly. A vigorous individual plunged into olive oil soon loses its brightness, but a feeble luminosity persists for a long time even after death.

The microscope shows us in the luminous substance of glowworms a cellular tissue, filled with what appears to be a soft yellowish grease; the whole is traversed by the trunks and branches of the tracheæ or air-tubes. This substance extends in a thin layer along the inner sides of the abdomen. It is this greasy substance, this *corps graisseux*, that Treviranus regarded as the source

* Comp. Anat.   † 'Manuel d'Entomologie.' Paris, 1828.

L

of the light of glowworms. According to Carus, this luminous organ " constantly receives a current of a liquid equivalent to blood." This flow, in his opinion, is the cause of the rhythmical character of the luminous emanation, as seen in *Lampyris italica*.

Guéneau de Montbelliard showed in 1782 (Nouv. Mem. de l'Acad. de Dijon, vol. ii. p. 80) that the eggs of *Lampyridæ*, which are small yellowish spheres, appear also phosphorescent in the dark. Carus has confirmed this observation, and assures us, at the same time, that the *larvæ* of these insects emit a greenish phosphoric light,—a fact which had been observed by Treviranus as early as 1802. Some authors have asserted that the chrysalis of the glowworm is slightly luminous.

Fig. 26.

Other *Coleoptera* are exceedingly phosphorescent. Such are the numerous species belonging to the genus *Elater*, known to the English as *Fire-flies*, and of which *E. noctilucus* (fig. 26) of

Latreille has been most attentively studied. This insect, of a dark brown colour, attains about one inch and a half in length. On its back are observed two smooth yellow spots. It is extremely common in the Antilles and the whole of South America, and emits a much more vivid light than the *Lampyridæ.* Besides the two yellow dorsal spots, which are very brilliant at night, there exist two others, hidden under the wing-cases or elytra, the light of which is only visible when the insect flies; it then shows four luminous points of great brilliancy. Moreover, the whole body of the insect appears glittering with light, which shines through the intervals existing between each segment or ring of the abdomen, and which is easily perceived when these segments are gently pulled asunder. The light which is emitted by the two points upon the thorax alone, is sufficiently strong to allow us to read by its aid the smallest print.

Under the common name of *Fire-flies* a great number of these exotic Elaters are indistinctly grouped. Most of them are smaller than *Elater noctilucus.* Closely allied to this is *E. ignitus.* But, besides these two species, Illiger has described ten others under the generic name of *Pyr phorus;* and Kirby and Spence state that seventy distinct species of these luminous insects are spread about the hot climates, from Chili to the south of the

United States. The species described by Illiger (in vol. i. of Annals of Nat. Hist. Soc. of Berlin) were taken in the Brazils, Peru, Buenos Ayres, Chili, Cuba, St. Domingo, and Guiana. At St. Domingo, where these insects are abundant, the inhabitants call them *Cucuij*. They are often applied to useful purposes, as lights, as decorations, and especially to destroy gnats which swarm in the dwellings. Considerable quantities are taken to this effect in the month of June.

On examining the luminous tissue in *Lampyris noctiluca, Elater noctilucus,* and *Elater ignitus,* Macartney observed that this tissue only differed by its yellow colour from the greasy intercellular substance which is found in other portions of the insect's body. In *E. noctilucus* and *E. ignitus* the light proceeds from masses of this substance closely applied underneath the transparent parts of the insect's skin. When the season for giving light is passed, this yellow tissue is absorbed and replaced by the ordinary intercellular tissue. In *Lampyris noctiluca,* besides this greasy tissue, the same author observed, in the last abdominal segment, two small oval sacs formed by an elastic spiral fibre, containing a soft yellow greasy substance of closer texture than that above-mentioned, and affording a more brilliant light than the rest of the luminous tissue.* The light emitted by

* Philosophical Transactions, 1810.

the yellow greasy tissue of the spots in the thorax of *E. noctilucus* can be communicated, it appears, to the interstitial tissue which pervades the whole of the insect's body. It was De Geer who first observed that light shone between the segments of the abdomen when these were separated one from another.

Morren, late Professor of Botany in the University of Liége, has studied minutely the structure of the luminous organ of *Lampyris noctiluca.** He has shown that the luminous sacs described by Macartney are connected with a multitude of trachean ramifications (air-tubes), and that the luminous property of the glowworm. appears to depend, to a considerable degree, upon the process of respiration. The trachean ramifications proceed from a large trachea which issues from a spiracle (breathing-hole) situated immediately at the side of the luminous mass, on the exterior of the insect's body. When this spiracle is closed the light is immediately extinguished, and reappears when the spiracle is opened. As insects have the power of opening or closing their spiracles at will, the glowworm can thus increase or diminish its light. This also explains why the light of the fire-flies (*Elater*) is more brilliant when the insects are flying, for then their spira-

* I have not read M. Morren's paper. It is abridged in Kirby and Spence's Introd. to Ent. p. 513 (edit. in one vol.).

cles are wide open, and respiration more energetic than when they are in repose.

An interesting observation was made formerly by Alexander von Humboldt. He drew out a very vivid light from an *Elater noctilucus* that was dying, by touching the ganglia of one of its anterior limbs with a piece of zinc and a piece of silver.

Some other *Coleoptera*, said to be phosphorescent, belong to the genus *Paussus*. Of this genus at least three species have been brought to Europe,—*Paussus lineatus*, from the Cape of Good Hope, *P. microcephalus* and *P. sphærocerus*, also from Africa. The habits of these insects are not well known. It is the species *P. sphærocerus* (fig. 27) which is stated to be phosphorescent. The

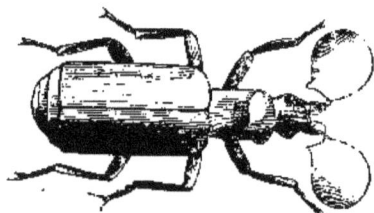

Fig. 27.

phosphoric light emanates from a peculiar swelling or vesiculous segment, which terminates the antennæ or horns of this curious insect. The fact was observed by Afzelius. Mr. Westwood, who has written a monograph upon this genus of insects, states that the "dim phosphoric light" mentioned by Professor Afzelius is probably only reflected light owed to the highly polished surface of the spherical cellular antennæ. But this is merely a supposition, and does not even amount to negative evidence. The genus *Paussus* is not mentioned in the twelfth edition of Linnæus's ' Systema Naturæ.' The species above named appear to be all night insects; they are easily recognized by the peculiar knob at the end of each of their antennæ, which are very short, and composed of only two segments, including the cellular one.

Lamarck thought that the two oval red spots which we remark upon the second segment of the abdomen of *Chiroscelis bifenestrata* are luminous in the dark. I am not aware that the fact has ever been confirmed. This insect, which is black, and about 1½ inch long, inhabits the Isle Maria. Its oval spots resemble those on the back of *Elater noctilucus.*

According to Latreille, the Chinese insect known as *Buprestis ocellata* has two spots upon its elytra, which are luminous at night. A friend of this naturalist assured him that he had seen these spots luminous in a living specimen which was brought from China to the Isle of France in some wood.

The *Scarabæus phosphoricus,* which Treviranus

speaks of as being luminous at the abdomen, is
no other than *Lampyris italica,* of which I have
spoken above.

One of the Longicorn beetles, named by Che-
vrolat *Dadoychus flavocinctus,* has the third and
fourth segments of the abdomen of the same yellow colour and appearance as the luminous segments of the *Lampyridæ,* and we know that a
considerable number of the Brazilian *Helopidæ*
present a similar peculiarity, whence some ento-
mologists are inclined to infer that these insects
are also luminous.

In the family of *Hemiptera* we have the genus
Fulgora, which includes several species said to be
highly phosphorescent. Their light is so brilliant
that the authors who speak of it have called these
insects *Lantern-flies.*

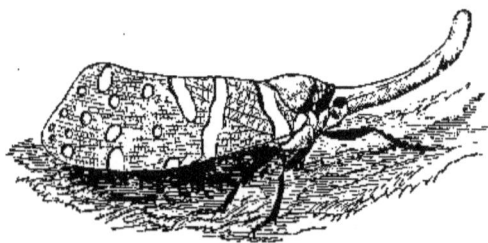

Fig. 28.

Fulgora laternaria and *F. candelaria* (fig. 28)
are the two species best known. These, like all
the insects of this genus, have a very singular ap-

pendage projecting from the head—a sort of long
proboscis. In *F. laternaria,* which inhabits South
America, this projection is horizontal, uneven, and
rounded at the extremity. In *F. candelaria,* which
is found in China, the proboscis is cylindrical and
curved upwards.

It is from these appendages, the sides of which
are transparent, that the phosphoric light emanates; they appear to be filled with a peculiar
phosphoric substance. Madame Mérian was the
first to observe the phenomenon. According to
her account, *F. laternaria,* which is a large insect,
emits a most brilliant light. It is said also that
the trunk of a tree covered with numerous individuals of *F. candelaria*—some in movement, others
in repose—presents a very grand spectacle, impossible to describe, but which may be witnessed
sometimes in China.

Fulgora pyrrhorynchus, described and figured
by Donovan in his 'Insects of India,' is said to
emit light of a fine purple colour.

Some authors have denied that the *Fulgora*
shine. Count Hoffmansegg informs us that his
insect collector, Sieber, a practised entomologist
of thirty years' standing, who took many specimens of *F. laternaria* in the Brazils, never saw
one luminous. On the other hand, the Marquis
Spinola, in the 'Annals of the Entomological Society of France,' vol. viii., contends for the lumi-

nous character of the whole tribe. Again, M.
Richard reared a species of *Fulgora,* but never
saw it shine; whilst a friend of M. Westmael assured him he had seen *F. laternaria* luminous when
alive.

Mr. Smith, of the British Museum, has related
to me an anecdote which confirms the opinion
that the *Fulgora* are certainly luminous. Whilst
curator of the Entomological Society, he was one
day showing some insects to Lady Seymour, her
son, a young midshipman, and one of his companions. The two latter had wandered to a case
of *Fulgora,* when one of the boys exclaimed,
" Why, look here! these are the *Candle-flies* that
we used to knock down with our caps in China."
Besides this, Dr. Donovan has carefully figured
these *Fulgora,* and his figures show them in the
act of emitting light from the points of their peculiar proboscis.

The group of *Fulgora* is, however, very little
known; we know scarcely anything of their
habits, except that they appear to be night-insects :
they merit assuredly a more complete history.

In the family of *Lepidoptera,* which includes
Butterflies, Moths, etc., a phosphoric light has
been observed in the eyes of *Noctua·psi,*—a little
grey nocturnal moth, which has upon its upper
wings a few black spots resembling the Greek
letter *psi.*

The same phenomenon appears to have been seen in the eyes of *Bombyx cossus.*

A Russian naturalist, M. Gimmerthal, has observed that the caterpillars of *Noctua occulta* are luminous, and the observation was communicated to the Entomological Society of France by M. Audouin. Since then, M. Boisduval remarked one hot summer evening in the month of June, a quantity of caterpillars on the stems of grass. They shed a phosphorescent light, and were certainly not the larvæ of *Noctua occulta,* but rather those of *Mamestra oleracea,* though they seemed larger than usual. M. Boisduval believes that this luminous appearance was caused by disease, which would account for its not having been met with before upon this common species.

If disease is capable of developing phosphoric light in insects, which it certainly has been known to do in superior animals, as we shall see further, it will account for the fact that other insects besides those named have been seen, though rarely, in a luminous condition whilst alive. Thus, some authors speak of *Grillus campestris,* or Mole-cricket, among the *Orthoptera,* as having been once seen in a phosphorescent state. Others assure us that the common " Daddy Longlegs," *Tipula oleracea,* was taken by a farmer who, seeing a light, knocked the luminous object down, and found it to be the insect just named.

Many insects which we are in the habit of observing only in the daytime may probably emit light in the evening, though all those which have been recognized with certainty to be highly phosphorescent, such as the *Lampyridæ* and the *Elateridæ*, are nocturnal insects.

CHAPTER VI.

PROBLEMATICAL CASES OF PHOSPHORESCENCE
IN SUPERIOR ANIMALS, AND PHOSPHORIC PHE-
NOMENA OBSERVED IN MAN.

IN the animal kingdom phosphorescence appears
to cease with the class of insects; in reality, however, phosphoric phenomena have been observed
in animals of superior organization during their
lifetime.

To speak, in the first place, of some problematical cases of phosphorescence in animals more
highly organized than insects, I will state that,
according to Carus, we are wrong in attributing
to a simple effect of reflected light that peculiar
scintillation which is observed in the eyes of dogs,
cats, tigers, etc. Rennger, in his 'Natural History of the Mammalia of Paraguay' ('Naturge-
schichte der Säugthiere von Paraguay'), published
at Basle in 1830, says, that he has seen the eyes of
a monkey so brilliant in complete darkness, that
they illuminated objects at a distance of half a
foot. The animal in question is the *Nyctipithecus*

trivirgatus, or *Singe Dormeur* (Sleeping Monkey), described for the first time by Humboldt, under the name of *Simia trivirgata,* in his ' Recueil d'Observations de Zoologie et d'Anatomie Comparée ' (vol. i. p. 306).

I have observed a brilliant scintillation in the eyes of man himself, but only once. The light was of a metallic-pink colour, resembling, in general aspect, the green light emitted from dogs' eyes. I only saw this in one individual, though I have examined many; but a friend of mine lately witnessed it in the eyes of a little girl. Both subjects alluded to were remarkably delicate.

I do not think we can attribute this light to phosphorescence, but rather, that it is owing to a phenomenon of reflection. I have never had an opportunity of ascertaining whether this luminosity of the eyes of human beings is visible in complete obscurity, as Rennger states was the case with the light emitted from the eyes of *Simia trivirgata;* but it is certain that the scintillation in the eyes of a cat or a dog is not visible in complete darkness.

In most cases it is not difficult to distinguish light which is reflected from light which is directly transmitted to us from the illuminated body itself, by means of the phenomena to which reflected light gives birth in the *polariscope,*—an ingenious

instrument imagined by François Arago. Now, in
the case of the phosphorescence of *Schistotega osmundacea,* spoken of in the second part of this
work, and those cases alluded to here, it would be
easy to ascertain whether the light was transmitted
directly from the plant or the animal itself, or owed
to the reflection of diffused daylight, as we observe
on the spiders'-webs in a semi-obscurity. Very
simple polariscopes, consisting of plates of crystal
inserted into flat pieces of cork, are sold by some
of the opticians of Paris ; whenever reflected light
is observed, at certain incidences, through this
simple apparatus, it shows coloured stripes, owing
to the polarization of the reflected light. With
directly transmitted light these coloured bands
are not visible.

A very remarkable case of phosphorescence was
witnessed by Dr. Kane in his last voyage to the
Polar regions, and described in his journal under
the date January 2nd, 1854. He was on his way
with Petersen to an Esquimaux settlement, in
order to procure food. Their thermometer was
at −42° Centigrade (−44° Fahr.). With their
weary dogs and sledge they had reached some un-
tenanted huts at a place called Anoatok, after
thirty miles' march from the ship :—" We took to
the best hut," says Dr. Kane, " filled in its broken
front with snow, housed our dogs, and crawled in
among them. It was too cold to sleep. Next

morning we broke down our door and tried the
dogs again. They could hardly stand. A gale
now set in from the south-west, obscuring the
moon and blowing very hard. We were forced
back into the hut; but after corking up all the
openings with snow and making a fire with our
Esquimaux lamp, we got up the temperature to
30° below zero Fahr.($-34\cdot5°$ Centigrade), cooked
coffee, and fed the dogs freely. This done, Petersen and myself, our clothing frozen stiff, fell asleep
through pure exhaustion; the wind outside blowing death to all that might be exposed to its influence. I do not know how long we slept, but
my admirable clothing kept me up. I was cold,
but far from dangerously so, and was in a fair way
of sleeping out a refreshing night, when Petersen
woke me with, 'Captain Kane, the lamp's out.'
I heard him with a thrill of horror.... Our only
hope was in relighting our lamp. Petersen, acting
by my directions, made several attempts to obtain
fire from a pocket-pistol; but his only tinder was
moss, and our heavily stone-roofed hut or cave
would not bear the concussion of a rammed wad.
By good luck I found a bit of tolerably dry paper,
and becoming apprehensive that Petersen would
waste our few percussion caps with his ineffectual
snappings, I determined to take the pistol myself.
It was so intensely dark that I had to grope for
it, and in so doing touched his hand. *At that*

instant the pistol became distinctly visible. A pale-bluish light, slightly tremulous, but not broken, co-vered the metallic parts of it—the barrel, lock, and trigger. The stock, too, was clearly discernible, as if by the reflected light; and *to the amazement of both of us, the thumb and two fingers with which Petersen was holding it,* the creases, wrinkles, and circuit of the nails, clearly defined upon the skin. *The phosphorescence was not unlike the ineffectual fire of the glowworm. As I took the pistol, my hand became illuminated also, and so did the powder-rubbed paper when I raised it against the muzzle.* The paper did not ignite at the first trial; *but the light from it continuing,* I was able to charge the pistol without difficulty, rolled up my paper into a cone, filled it with moss sprinkled over with powder, and held it in my hand whilst I fired. This time I succeeded in producing flame, and we saw no more of the phosphorescence. . . . Our fur-clothing, and the state of the atmosphere, may refer it plausibly enough to our electrical condition."*

I have given this account in Dr. Kane's words, that the conditions under which this curious production of light occurred may be more readily appreciated.

The light arising from currying a horse, or rubbing a cat's back, has often been observed. A

* See Dr. Kane's work, 'The Second Grinell Expedition,' etc.

peculiar snapping noise accompanies this light, which is owing simply to the production of a quantity of electric sparks. Similar instances of spontaneous production of light have been observed on combing a woman's hair, provided it be very dry, and the atmosphere devoid of moisture.*

Bartholin gives an account of a lady in Italy, whom he designates as *" mulier splendens,"* whose body shone with phosphoric radiations when slightly rubbed with a piece of dry linen.

A phenomenon, which might perhaps be termed *subjective phosphorescence,* occurs when injuries are received by the eye or the optic nerve (fig. 29).

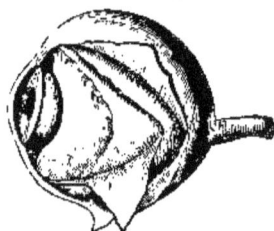

Fig. 29.†

Thus, when the head has been held down for some moments, sparks are often seen before the eyes on resuming an upright position. These sparks

---

* I have observed this frequently, but generally when a south-west wind is blowing, and the atmosphere highly electrical.

† Fig. 29, the human eye, showing the different membranes and the optic nerve.

appear in motion, and are of frequent occurrence in some diseases, such as typhus fever, when the invalid sees them upon the bed-cover, and constantly endeavours ˙ to pick them off with his fingers. When a blow is received upon the eye, an intense light is perceived : this must be familiar to pugilists.

When the optic nerve is cut, no pain is felt; but an intense flash of light across the eyes is experienced. The same flash occurs when an electric current passes through the optic nerve, as sometimes happens when a piece of silver and a piece of zinc are made to form a galvanic couple with the tongue or other parts of the mouth. To en sure the success of this experiment, the plates of zinc and silver should be placed upon the inside of each cheek, and connected together, outside the mouth, with a piece of silver wire.

I have observed very vivid luminous appearances, during fever, in my own eyes; they manifested themselves after any violent exertion, such as going upstairs, walking quickly across a room, etc.* The light appeared in the form of myriads of luminous spots, in rapid motion, and of a greenish-yellow tint.

These phenomena, to which I have given the

---

* These actions would not amount to *violent* in a state of health, but require a great amount of exertion during illness.

name of *subjective phosphorescence,* were first accurately described by Purkinje.

Ritter, Purkinje, and Hjort have noticed that when the eye is included in a galvanic current, a vivid flash of light is perceived whenever this circuit is closed or opened. Purkinje also remarked that whenever the eye is pressed with the finger certain arborescent figures are produced, all of which are luminous.

Narcotic medicines often affect the eyes in a similar manner, producing subjective phosphoric radiations. A production of light in the above circumstances is exceedingly interesting, and tends, perhaps more than we are aware, to establish the fact that the phenomena of light are owing to a *vibratory movement of matter.*

In a former chapter I have spoken of the appearance of phosphorescent light upon the bodies of animals, and upon the human body after death. In this one I have to mention similar appearances upon the living body.

Marsh, in an 'Essay upon the Evolution of Light from the Human Subject,' brings forward the following statement made to him in these words:—"About an hour and a half *before* my sister's death, we were struck by luminous appearances proceeding from her head in a diagonal direction. She was at the time in a half recumbent position and perfectly tranquil. The *light*

was pale as the moon, but quite evident to mamma,
myself, and sisters, who were watching over her
at the time.   One of us, at first, thought that it
was lightning, till shortly afterwards we fancied
we perceived a sort of *tremulous glimmer* playing
around the head of the bed; and then, recollect-
ing that we had read something of a similar
nature having been observed previous to dissolution, we had candles brought into the room,
fearing our dear sister would perceive it, and
that it might disturb the tranquillity of her last
moments."

We are told of a similar luminous apparition
around the person, and in the room, of a man
who had been lying ill of a lingering disease, of
which he afterwards died, in the south-west of
Ireland.

In 1840, Donovan published, in the 'Dublin
Medical Press,' a very curious case of phosphorescence upon the living body of a man.   "I was
sent for," he says, " to see Harrington, in December, 1828.   He had been under the care of my
predecessor, and had been entered in the dispen-
sary book as a phthisical patient; and on referring
to my note-book, I find that the stethoscopic
and other indications of phthisis were indubit-
able.   He was under my care for about five years,
during which time the symptoms continued stationary, and I had discontinued my attendance

for about two years, when the report became general that *mysterious lights* were seen every night in his cabin. The subject attracted a great deal of attention. . . . I determined to submit the matter to the ordeal of my own senses; and for this purpose visited the cabin for fourteen nights. On three nights only did I witness anything unusual. Once I perceived a *luminous fog*, resembling the Aurora Borealis; and twice I saw *scintillations, like the sparkling phosphorescence exhibited by sea-infusoria.* From the close scrutiny I made, I can with certainty say that no imposition was either employed or attempted."

These strange luminous apparitions are never seen but in cases of extensive disease. The theories that have hitherto been brought forward to explain them are quite inadequate to account for these phenomena.

I read also of another similar case of phosphoric light glimmering about the bed of a woman in Milan. This light fled from the hand which approached it, and was at length entirely dispersed by a current of air.

Of the same kind are those luminous appearances which are sometimes, though rarely, seen in houses, and which have been called "*Elf-candles*" by the Scotch. They are supposed to portend the death of some person in the house. As they have been known to occur *before* death upon the *living* subject, I have separated them from those mentioned in the first chapter of this part of my work, although they may probably be referable to the same cause.*

---

* It is by no means proved that death inevitably follows this phosphorescent light: the case of the Milanese woman seems to prove the contrary.

PART IV.

HISTORICAL, THEORETICAL, AND PRACTICAL CONSIDERATIONS.

CHAPTER I.

HISTORICAL NOTES.

I HAVE now exposed all the cases of phosphorescence with which I am acquainted, in minerals, mineral substances artificially produced, in vegetables, both phanerogams and cryptogams, and in animals dead or living, besides having alluded to several singular cases of meteorological, and some problematical cases of astronomical, emission of phosphoric light.

It will thus be seen that phosphorescence pervades the whole of nature. Not only have spontaneous evolutions of light been witnessed in our chemical and physical laboratories as curiosities produced by human skill, we find the same interesting phenomena exhibited in natural mineral substances, in various plants and animals, where our physical knowledge takes no part in their production. Moreover, we see phosphoric light developed in the atmosphere, as, for instance, luminous fogs, and in the heavens.

We thus become impressed with the generality of these phenomena, which have hitherto been regarded only as rare and mysterious evolutions of light.

It is curious to note the progress of the discovery of phosphoric phenomena, and the ineffectual attempts that have been made to explain them as our knowledge gradually extended.

The ancient Greeks gave the name Phosphorus to the *morning star,* or planet Venus, when it rises before the sun. By the Latins, the name of Lucifer was given to this star, which the French term *Etoile de Berger.*

Although the phosphorescence of the Bologna stone, and certain other *phosphori,* was only observed in the seventeenth century, the luminosity of the sea was familiar to observers in the darkest ages of antiquity. Some authors assert that it was generally attributed by the ancients to Castor and Pollux; but the phenomenon attributed to these divinities was simply the electric light which I have alluded to as the *Fire of St. Elmo,* appearing in stormy weather on the masts of ships.

Aristotle mentions light proceeding from putrescent substances and from glowworms; Pliny was acquainted with the luminous properties of the dead *Pholas* and certain *Medusæ,* and this ancient naturalist knew that by rubbing one of these animals upon a plank the wood became luminous, and that when it had ceased to shine, the light could be brought back again by rubbing the hand over the board.

In 1592, Fabricio d'Aquapendente brought forward the first observation of phosphorescent flesh, which Robert Boyle afterwards attempted so energetically to explain.

In 1602, Cascariolo discovered the famous Bologna stone, in the manner already described.

In 1663, Robert Boyle found that diamonds possessed the properties of solar-phosphorus.

In 1669, Brandt discoverd the chemical element phosphorus, re-discovered by Kunckel in 1674, and, according to some, by Robert Boyle in 1680. Boyle's account of this new substance was indeed only published in 1680. Some authors say that he saw phosphorus in the hands of Kraft, and knew that it was extracted from some matter pertaining to the human body; others say that Brandt had partially communicated his secret to Kunckel. However this may be, the discovery, after all, was made by Brandt.

The remarkable light emitted by phosphorus in the dark, known at present to be an effect of oxidation, was supposed to be owing to combustion; and strange to say, this opinion, which was not till long afterwards confirmed by direct experiment, was the exact representation of the phenomenon.

In 1686, Tachard, an ecclesiastic, in order to explain the phosphorescence of the sea, stated that the waters of the ocean absorbed the light of the sun by day and emitted it again at night. The distinguished philosopher Robert Boyle, who lived at this period, believed that the light of the waves was owing to friction. He imagined that the atmosphere rubbed against the water of the sea by the rotation of the earth, and that this friction had for direct effect the emission of a certain amount of caloric and light. We have already seen that Boyle endeavoured also, by numerous and ingenious experiments, to account for the phosphorescence of rotten wood, flesh, etc.

Later still, we find that Mayer reproduces the old opinion of Tachard; and Beccaria affirms that the solar-phosphorus, or Bologna stone, "absorbs light, and emits it some time afterwards." Beccaria thought he had observed that this substance, when submitted to coloured light from red, yellow, blue, and green glasses, shone in the dark with a red, yellow, blue, or green light. But this was afterwards distinctly denied by Wilson in England, Zanetti and Algerotti in Italy, by Dufay in France, and by Grosser of Vienna.

In 1797, Brugnatelli published a singular opinion in the 'Annali di Chimica.' He believed that the phosphorescence of the *Lampyridæ*, or glowworms, was owing to a peculiar physiological

act, by which these insects separated light from
their food, and afterwards secreted it in a sensible
form. When we recall the chemical theories in
vogue at that period, we find that Brugnatelli's
opinion is not so ridiculous as one would be apt
to suppose.

Carradori, another Italian naturalist, appears
to have admitted Brugnatelli's opinion; but know-
ing that the *Lampyridæ* could extinguish or emit
their light at will, he thought that they effected
this with a peculiar membrane acting as a screen,
by which the insect (*Lampyris italica*) hid its
light. The existence of this membrane was afterwards denied by Macartney in the 'Philosophical
Transactions for 1810.' But Carradori caused
science to take a step forward when he proved
that the light of *Lampyris italica* was not immediately extinguished *in vacuo*, in oil, or under
water, as the light of a candle, or that of common
phosphorus would be.

Boyle, Hume, and Macaire have all observed
that the phosphorescence of certain dead organic
matters was extinguished, at least partially, *in
vacuo*. Boyle's experiments on this subject were
published in the 'Philosophical Transactions' for
1668. Although he certainly did happen once to
extinguish the light of phosphorescent wood, *in
vacuo*, completely, he never succeeded in totally
extinguishing that emitted by dead fish.

Experiments, the results of which appear contradictory, were made upon the glowworm's light, by Forster, Spallanzani, Dr. Hulme, Beckerheim, and Humphry Davy. Davy's experiments show that the light was neither increased in oxygen or extinguished in hydrogen gas (Phil. Trans. 1810). It was in 1749 and 1750, as we have seen, that Professor Viannelli and Dr. Grixellini discovered, in the Adriatic Sea, their small luminous worm, *Nereis noctiluca;* and from this moment the real cause of the phosphorescence of the sea was established. It was then owing to animalcules! Soon afterwards Captain Cook and Mr. Forster met with those curious little organisms, the *Noctilucæ,* and recognized them as the cause of the phosphorescence of the ocean. In 1776, the Abbé Spallanzani treated of the phosphoric light emitted by *Medusæ.*

Since then, discoveries connected with phosphorescence have multiplied considerably up to the present day.

The history of that of our own *Noctiluca miliaris,* which illuminates the waters of the English Channel, and therefore interests us particularly, is curious enough.

This animalcule was accurately observed, for the first time, by a French naturalist, Rigaud, in 1765, who speaks of it in the 'Mémoires de l'Académie de Paris;' it was seen also, about the same time, by Slabber of Harlem. In 1775, Dicquemare discovered it again in the sea at Havre, and in 1810, M. Suriray called attention to its existence upon the same coast as a novelty. He gave it the name *Noctiluca miliaris,* which it has since retained. In 1834, Professor Ehrenberg, whilst studying the phosphorescence of the sea on the coast of Heligoland, met with the same little animal, and called it *Mammaria scintillans,* by which name it is designated in some of Humboldt's works. M. de Quatrefages, of Paris, has written an interesting paper upon it; and lately, in 1855, Dr. Verhaegho, of Ostend, published a small pamphlet, 'De la Phosphorescence de la Mer sur les Côtes d'Ostende,' containing the results of his observations upon this curious little being.

The dates of other important discoveries connected with phosphorescent phenomena have been given in preceding chapters.

In 1775, Wilson discovered that the most refrangible rays only of the solar-spectrum acted upon the different kinds of *solar phosphori;* Beccaria, from his own experiments, came to the same conclusion about the same time. In 1802, Engefield re-discovered that the blue rays of the solar-spectrum acted more energetically than any others in promoting the phosphorescence of the Bologna stone, and his experiments were repeated and confirmed by Ritter, Goethe, and Seebeck.

Some papers upon phosphorescence after insolation, were afterwards published by M. Dessaignes; they gained the prize of the Academy of Sciences, at Paris, in 1807–1808. This author was the first to observe that *substances which are bad conductors of electricity* are very easily rendered phosphorescent, whilst those which are *good conductors* apparently very rarely or never.

Dessaignes, moreover, remarked that electricity, either in the shape of an electric spark or discharge, or in that of a simple current devoid of light, gives the faculty of shining in the dark to bodies which do not appear to possess it.[*] Dessaignes was therefore naturally brought to the conclusion that every case of phosphorescence is intimately connected with electricity.

Some years later, M. Becquerel and M. Biot, in France, and afterwards, Professor Henry, in America, repeating the experiments of Dessaignes, and adding some new ones of their own, arrived at the same conclusion.

But they were not the only observers who were occupied upon this subject; whilst Grothuss, in Germany, who seems to have been the first to recognize the action of an electric discharge upon phosphoric bodies, was evidently remaining be-

---

[*] Grothuss seems to have been the first to show that the diamond became luminous when an electric discharge was passed over it.

hind, and believed in the old opinion of Beccaria, Heinrich, at Nuremberg, and Pearsall, in London, supported the ideas of Dessaignes. Becquerel and Biot have the merit of having first studied the influence of *transparent screens* of different substances (glass, quartz, calcareous spar, etc.) in promoting or extinguishing phosphorescent light produced by insolation. Their experiments were repeated recently by Professor Henry, of Philadelphia. The action of *coloured* glasses upon this phenomenon had already been investigated by Wilson and Beccaria.

I shall not repeat here what I have stated in other parts of this work concerning the discovery of plant-phosphorescence and of light-emitting animals.

For the benefit of those who might desire to consult some of the more important and original documents relating to the subject of phosphorescence, I have determined to give, in an Appendix to this work, their titles, and the dates of their publication.

As we have already seen, the light emitted by flowers is thought to be owing to electricity; but as for that which is evolved from *Fungi*, such as the *Agaric of the olive*, the *Rhizomorpha*, etc., no idea has been formed of the direct cause of the luminous phenomenon.

According to Matteucci, Roberts, and De

Quatrefages, the light of *glowworms* is a phenomenon of combustion—such, at least, was the
opinion of these philosophers when their observations were published.

According to De Quatrefages and Professor
Ehrenberg, the light of the *Noctiluca* is an electric
phenomenon. Ehrenberg has discovered in a little
marine worm, *Photocharis cyrrigera*, that which
appears to him the *organ* of *phosphorescence.* He
thinks also, that those small elevations which we
observe upon the bodies of *Noctiluca,* when highly
magnified, are so many *phosphoric organs;* so
that it seems evident, at present, that wherever
phosphorescence is observed in animals, there
exists a special organ destined to fulfil this func-
tion.

CHAPTER II.

THEORY.

It will be easily seen, by examples that have been brought forward in preceding chapters, how difficult and how ineffectual have been the attempts to give a satisfactory explanation of phosphorescence. The opinions on this subject published by Dessaignes and Becquerel certainly merit the most consideration. The theory to which they lead may be seen condensed in the following paragraph :—

"It is perfectly demonstrated, at the present day, that an evolution of electricity takes place in bodies whenever the equilibrium of their molecules undergoes a change of any sort, either in their chemical constitution or in their physical aggregation. If these molecules are not separated thereby, we observe a recomposition, more or less rapid, of the two electricities that, for an instant, were put in liberty; and this may determine, according to the nature of the body and the tension of the electricity, a production of light and heat. Hence, when the particles of a body are shaken by percussion, friction, heat, light, or decomposed by chemical action or by an electric spark, these two effects (light and heat) may be produced by the recomposition of the two electricities, especially when the particles submitted to experiment are *bad conductors*. But as these causes are precisely those which produce phosphorescence, we are induced to admit the identity of electric light and that of phosphorescence;* so much the more as the luminous appearances are sensibly the same in both cases, and as bodies, which are *good conductors of electricity*, in which the phenomena are rarely accompanied by emission of light, are also devoid of phosphorescence" (*Becquerel*).

This theory, brought forward some years ago, is hardly on a level with the present state of science; and, indeed, Dr. Young's ideas on the phosphorescence of solar phosphorus, appear to me quite as near a satisfactory explanation. Dr. Young admitted that the shining of the Bologna stone, after it has received the rays of the sun, greatly resembles the sympathetic sounds of musical instruments, which are agitated by other sounds conveyed to them through the air.

* It would not be impossible to prove this opinion to be true or false, by submitting phosphoric light to spectrum analysis, as I have stated before. For my own part, I am not inclined to admit the fact *à priori*, in M. Becquerel's sense.

All the physical researches of times gone by, and all the experimental data furnished by the philosophers of the present day, tend to prove the reality of those admirable notions brought forward by Grove in his ' Correlation of Physical Forces.' The correlation theory enables us to account for many hitherto unexplained phenomena, and points out to us the direction in which physical science may spread to greatest advantage.

I am indeed led to believe that all those re- markable phenomena classified under the heads of light, heat, electricity, magnetism, etc., are in reality *modes of motion,* or matter in motion. When two bodies of a certain volume move before us, we can witness and describe the motion easily enough ; but when motion takes place among the molecules of bodies, the most powerful microscope will not allow us to detect it : we are led, how- ever, by innumerable facts, to infer that such motion does occur. By stating that the so-called " physical forces " are different modes of motion, I understand that if the molecules of any substance vibrate in one direction, north-south for instance, we have *light;* in another direction, east-west, *electricity;* in an intermediate direction, *heat;* in another direction, *magnetism,* etc. All these mo- tions being connected to the primary rotation of our planetary system, or *gravitation.* But we have no proof that the molecules of bodies vibrate in straight lines; their motion is more probably circular. Indeed, my ingenious friend M. Porro has endeavoured to show the great resemblance which seems to exist between these molecular movements and those of celestial bodies; and it has been supposed by some philosophers that the molecules of matter are as distant from each other, in proportion to their size, as the planets themselves.

But, in the present state of knowledge, all these considerations are premature.

We must not confound the luminosity produced in substances that are heated to a certain degree which is supposed to be identical, or nearly so, for all, with that peculiar *phosphoric radiation* which a great number of substances emit at different degrees of heat, and some, at the ordinary temperature of summer. As early as 1776, Dr. Fordyce observed, that heated bodies began to be luminous in the dark at from 600° to 700° of Fahrenheit's thermometer (see his interesting paper in the 'Philosophical Transactions' for 1776).

I say that we must not confound these two manifestations of light, though they are both molecular vibrations of the body submitted to experiment; yet they differ: for instance, fluor-spar heated gently over a fire becomes phosphorescent; heated to a still higher degree, the

phosphorescence disappears, and is afterwards replaced by another kind of light; the body is then
said to be *red-hot, white-hot,* etc.

In every case phosphoric light is developed
first, and is followed by the second kind of luminosity, unless chemical decomposition is required
to take place before phosphorescence ensues.*

These two kinds of light have never been examined physically to ascertain whether they differ
in any essential property, but we know that they
must differ. Researches undertaken with this
view would be exceedingly interesting. We know
that the light of incandescent solid bodies, that is,
bodies heated red-hot, and the light of the electric
spark, exhibit great diversity in the number and
position of Wollaston's dark lines, already referred
to in this work. The velocity of electric light is
also known to be greater than solar light in the
ratio of three to two, according to Wheatstone's
admirable experiments. Again, we know that
a solid or liquid incandescent body possesses light

* For instance, when native gypsum is heated on charcoal
before the blowpipe, some curious phenomena occur. After the
compound has lost its water, it melts into a beautiful transparent
bead, which becomes opaque on cooling. If this be strongly
heated in the flame of reduction, the sulphate of lime loses its
oxygen; at the same time, its point of fusion is considerably
raised, and it is transformed into sulphuret of calcium. At this
moment, with greater heat, it begins to melt again, and at the
same time a very brilliant phosphorescence is observed.

which differs in certain properties, from that produced by a burning or luminous gas, as shown by Arago; the former gives indications of polarization when viewed at an acute angle, whilst the latter never shows any traces of polarization at whatever angle it is viewed.

All these experiments remain to be made with *phosphorescent bodies.* We are therefore very much in the dark as to the nature of phosphoric light.

We are so accustomed to associate *light* and *heat,* as in the flame of a candle, for instance, that it is difficult to bring the mind to reflect upon a fact beyond doubt, namely, that the one may be generated without the other. When a wire becomes heated by an electric current, it often becomes luminous also; but in other cases a body may become luminous without any sensible degree of heat. With combustible substances used in candles, lamps, etc., the *greater the light the less the heat, and reciprocally.* The flame of an oil lamp is very hot; that of a camphine lamp, which is far brighter, is very much cooler; whilst the flame of a spirit lamp, quite invisible in the sunshine, produces a great amount of heat. Let us inquire how these things occur.

We know that whenever any one of those specific motions of matter which we term *force* (light, heat, electricity, etc.) ceases to manifest itself, it is replaced immediately by another, equivalent for equivalent. Thus, a *force* A, being given in action, as soon as it ceases to act, we see it replaced immediately by its equivalent of another force B. And when the force B ceases to manifest itself, it is immediately replaced by its equivalent of another force C, D, or E, or the force A re-appears.

Hence we see one kind of motion (friction) transformed into *heat* or *electricity,* according to the substance submitted to experiment. By rubbing wood we produce heat, by rubbing glass or resin, electricity. Again, motion (pression) is transformed into *electricity,* when we press the angles of certain crystals, such as Iceland spar. In the same way *heat* can be transformed into *motion* (steam-engine), into *electricity* (thermoelectric currents), into *light,* into *chemical action,* etc. And each of these new kinds of motions (forces) generated may in its turn be transformed into *heat* or into any of the others. By heating water, the molecules of this body are put into *motion;* by heating a bar of antimony soldered to a bar of bismuth, a certain amount of the vibration called *heat* is transformed into its equivalent of *electricity.* The same transformation takes place when crystals of Tourmaline or Boracite are heated.

When we heat a platinum wire by means of an

electric current, we see a certain amount of *electric vibration* transformed into *heat* and *light,* for the wire first becomes hot and then luminous. If we place upon the snow two pieces of cloth—one white, the other black—the white one reflects back the light of the sun, the dark one absorbs the light, and a certain amount of it is transformed immediately into *heat,* which, in its turn, is transformed into another kind of *motion* manifested by the melting of the snow under the black cloth.

In the same way *light* is transformed into *chemical action* (as in photographic processes), and *chemical action* into *light* (combustion of phosphorus). So, again, the electricity of the galvanic battery is transformed into its equivalent of *light, chemical action, heat,* etc., and *chemical action* gives birth to *electricity, light,* and *heat.*

The motion or vibration of a special organic substance, which motion constitutes *nervous force,* can be transformed into *electricity,* and *electricity* into *nervous force.* Matteucci, who has made numerous experiments upon this subject, distinctly insists upon the necessity of distinguishing between electric force and nervous force. One of these can give birth to the other, and reciprocally; but they are not identical. When a nerve is excited by an electric current, it immediately occasions muscular contractions; but at this moment not a trace of electricity can be detected in any

part of the nerve. As soon as electricity meets
with the nerve, it finds the necessary medium for
transforming itself into another kind of vibration,
which we call nervous force, and this passes into
muscular motion, or contraction. Reciprocally,
when we hold in our hands the two wires of a
galvanometer, and, by a muscular contraction,
set the needle of the instrument in motion, we
cannot say it is nervous force which moves the
needle; this motion is owing to *electricity,* result-
ing itself from the transformation of a certain
amount of *nervous force.*

It is highly probable that this doctrine of transformation may apply to *nervous force* and *instinct*
or *will;* though we enter here into considerations
which are beyond our present means of experiment.

In nature, we can almost always connect *light*
with electricity as a starting point,—especially
when it concerns *bad conductors.* When an electric current passes through a bad conductor, a
great amount of electricity is transformed into
light, and the body experimented on becomes
luminous. Again, in the *combustion* of phosphorus, as in every chemical action, a certain
amount of what we call *chemical force* or *chemical
affinity,* is transformed into *electricity;* and, in the
case of phosphorus and many other bodies, a
portion of this electricity into light. This latter

transformation depends upon the nature of the body and the intensity of the action.

The phosphorescence of minerals, or mineral substances artificially produced, is an example of one of the vibrations of matter already alluded to, and which may be owed, in the first place, to heat, electricity, solar light, chemical action, etc. In a great many cases, as Dessaignes and Becquerel have shown, electricity is the immediate vibration to which the light produced may be referred; and that is the reason why *bad conductors* are more readily phosphorescent than other bodies, and, probably, why the most refrangible rays of the solar-spectrum are the only ones which will induce phosphorescence after insolation.

My own idea of *phosphorescence after insolation* is as follows:—The light of the sun, acting upon a mineral substance, occasions a certain vibration (electric, chemical, or magnetic); but this vibration not being able to continue when the action of light ceases, that is, when the substance is placed in obscurity, the body gives back light whilst losing the vibration (electric, chemical, or magnetic) occasioned in it by the rays of the sun. *The body in question does not, in this case, give back the entire quantity of light it has received; but a quantity equivalent to the electric, chemical, or magnetic vibration induced in it by the direct influence of the solar light.*

The reason why I suppose a *magnetic* vibration may be induced as well as a chemical or electric motion, is obvious, since Morichini and Mrs. M. Somerville have shown, by direct experiment, that magnetism is induced in steel by exposure to the most refrangible of the solar rays, precisely those which induce phosphorescence.* Moreover, light is engendered by magnetism in a curious experiment made by Grove in 1845. A tube, filled with the liquid in which magnetic oxide of iron has been prepared, and terminated at each end by plates of glass, is surrounded by a coil of coated wire. " To a spectator, looking through this tube, a flash of light is perceptible whenever the coil is electrized ; and less light is transmitted when the electrical current ceases, showing a symmetrical arrangement of the minute particles of magnetic oxide while under the magnetic influence." (Correlat. of Phys. Forces, p. 158.) Some remarkable results were likewise arrived at by Sir W. Snow Harris, as early as 1834, by vibrating magnets in the sun, and published in the ' Edinburgh Philosophical Transactions' for that year.

A similar conception would apply to phosphorescence produced by heat, chemical action, electricity, etc.

---

* These are also the rays which cause chlorine to combine with hydrogen, and which decompose many chemical compounds.

It is well known that when a platinum wire is suspended, after being heated, in a mixture of ether-vapour and air, or in a mixture of alcholic vapour and air, it continues incandescent for hours together, until all the ether or alcohol employed is spent. In this case, the spirituous vapour is burnt by the oxygen of the air; but neither the oxygen nor the ether become luminous,—it is the wire alone which gives out light. This curious phenomenon may go far, perhaps, to explain the phosphorescence of dead animal matter. The whole circumstances connected with it have been brought forward in my prize memoir, 'La Force Catalytique, ou Étude sur les Phénomènes de Contact,' printed at Harlem in 1858 (see Appendix).

As regards phosphorescence in the vegetable kingdom, possessing, as we do, only a few isolated observations, we are devoid of sufficient experimental data to offer any theoretical consideration as to its production. If the light emitted by flowers appears to be electrical, that evolved from fungi is more probably connected with chemical action.

In luminous animals, we find everything prepared by nature for the production of light; namely, a phosphorescent organ specially adapted for this purpose.

Taken in its *most satisfactory* point of view, the

evidence of combustion being the cause of the
light of glowworms may be thus stated :—

1. Matteucci and Roberts, the former by very
delicate experiments, assure themselves that a
slow combustion takes place, though no sensible
heat is evolved. They do not know what substance burns; they find, however, that the light,
though not immediately extinguished in hydrogen
and carbonic acid, is so in about half an hour;
whilst, in oxygen gas, it shines three times longer :
that is, at least an hour and a half. 2. M. Schnetzler finds *phosphorus* in the luminous tissue of the
insect, *after oxidizing it* with nitric acid, in the
state of *phosphoric acid* or *phosphates;* and this
phosphorus *may* have been present as *free phosphorus,* since Mr. Thornton Herapath finds no
phosphates in the insect's body. 3. Professor
Morren, as Macartney had done before him,
shows that large air-tubes or tracheæ are intimately connected with the luminous tissue; and
Morren shows further that the animal extinguishes
its light by closing the spiracula or air-orifice
through which the air enters the luminous organ.
4. The luminous substance shines for some time
after death, as if phosphorus were really present,
especially when damp.

Until these facts, which tend to prove that the
phosphorescence of glowworms is a phenomenon
of combustion, be confirmed or refuted by fur-

ther research, I cannot do otherwise than repro-
duce here the theory I published for the first time
in 1858.

In the glowworm, the luminosity can be traced
directly to the instinct of the insect through what
are termed the correlative forces, electricity and
nervous force.   We find the phosphorescent
organ of a greasy nature, a bad conductor of electricity, under the dependence of the nerves, which
in their turn depend upon the instinct.   In the
Fireflies there exist similar organs, destined by
nature to produce light ; and, as anatomical science
progresses, the same will doubtless be found in
those myriads of inferior organisms endowed with
phosphorescent properties.   Indeed, Ehrenberg
has already described what he takes to be the
phosphoric organ in *Noctiluca miliaris* and *Photocharis cyrrigera.*

A given amount of electricity will always produce an equivalent proportion of light when
passing through a bad conductor ; and a certain
amount of nervous force, acting through the
nerves, is capable of producing an equivalent
amount of electricity ; finally, it is doubtless true
that instinct is correlative with what are called
the other modes of force.

It will be objected, perhaps, that the luminous
substance extracted from the body of a *Lampyris*
shines for some time after the death of the insect ;

but can we not contract the leg of a frog long after death by means of a galvanic current? What is termed *nervous force* exists, then, for some time after death, as numerous observations show; and when *it* disappears, no more contractions— no more light.

.

CHAPTER III.

PRACTICAL CONSIDERATIONS.

To those reflective minds who really enjoy the beauties of nature and contemplate them with admiration, practical details are of so little interest, that I had not the intention of saying anything upon the subject in this work. But, as some persons endeavour, at any risk, to turn everything to profit, to make everything in nature useful to man in one way or another, I am willing to show that phosphorescence is far from being devoid of utility, and that it has already been extensively applied to various uses, often without a knowledge of the fact.

It would, indeed, be superfluous here to dilate upon the uses of *light;* it will be sufficient for me to state that the most powerful light ever produced by man, without exception, is owed to a phenomenon of phosphorescence. I allude to the "Drummond, or Lime-light."

It is produced by submitting lime, a highly

phosphorescent substance, to the heat, or, rather, to the chemical and electrical action, engendered during the combustion of hydrogen by oxygen gas. I have indeed been able to show by direct experiment that *heat* has not so much to do with the production of this intense light as is generally supposed.

If a piece of borax be heated before the blow-pipe, it melts when it has attained a red heat. If it be now allowed to cool and a little lime sprinkled over it, on applying heat a second time, the lime becomes vividly phosphorescent long before the borax is at all affected by the heat. It is therefore evident that lime glows vividly with phosphoric light long before its temperature has attained what is termed a red-heat.

In the case of the Drummond Light and the phosphorescence of lime before the blowpipe, electricity may probably be the force which induces phosphorescence, since Grove (in 1854) has shown that the blowpipe flame gives rise to a very marked electric current; a number of jets actually enabled the author to form an electric battery of a certain intensity. Now, when the flame of oxygen and hydrogen gases is employed, the electric action must naturally be far more considerable than with a simple blowpipe flame. Hence the vivid phosphoric light which bears the name of Mr. Drummond, and which has been applied to so many useful purposes by engineers, astronomers, microscopical observers, etc.

The same arguments might apply to the Electric Light; but it has not yet been proved that this is identical with phosphorescence, though such is the opinion of some philosophers, as we have already seen.

Again, I have alluded, in the first part of this work, to the fact that houses freshly painted with lime-wash are frequently luminous at night, though slightly, after exposure to the sun's rays during the day. If, by chemical and physical research, a means were discovered capable of rendering this phosphoric light more powerful, by employing sulphides of calcium or barium, etc., and superadding, if necessary, the action of an electric current when the sun is hidden by clouds, a street might be effectively illuminated by phosphorescent light alone.

The light of the *Elateridæ*, or Fireflies, has for years been employed for lighting apartments, and in travelling, in the West Indies.

Phosphorus dissolved in oil has been sometimes used as a night-light. As long as the bottle which contains the liquid remains closed no light is seen ; but when opened the phosphorescence is sufficiently bright to enable us to read the hour upon a watch, etc. These luminous bottles were once much in vogue as useful curiosities.

In cases of poisoning by phosphorus, the luminosity of this substance in the dark is the principal character by which chemists assure themselves of its presence in the intestines, etc., submitted to analysis. Mitscherlich, of Berlin, has invented an ingenious apparatus for this purpose.

If we were acquainted with the circumstances which produced the phosphoric light described by Dr. Kane, and which enabled him to find the pistol so readily, it would be exceedingly useful to command at will such an evolution of light.

As concerns Natural History, new sources of light employed in microscopical investigations are often attended with unexpected results. Such was the case when polarized light was first applied in this sense; and we know that the solar microscope may be brought into action at night by means of the phosphoric light evolved by lime. I have known phosphorescence had recourse to in order to distinguish one plant from another. This was the case with two closely allied species of *Rhizomorpha*, which sometimes resemble each other extremely; the one is however phosphorescent at night, and the other devoid of this property.

In Mineralogy the phosphorescent properties of fluor-spar, arsenic, lime, oxide of zinc, etc., render it easy to detect their presence when substances are heated before the blowpipe.

Again, no stone resembles the diamond so closely as white topaz, which is often sold as diamond. But the phosphorescent properties of the latter furnish us with a ready means of detecting the one from the other.

The reproduction of engravings, etc., by phosphorescence has been achieved by my friend M. Nièpce de St. Victor, as stated in Chapter VII. (Part I.).

As phenomena of phosphorescence are more studied, they will doubtless lead to new views upon the molecular constitution of bodies, which cannot fail to be attended with practical results.

Phosphorescence will certainly be applied to many other useful purposes as it becomes better known. We must remember, however, that it is yet in its infancy, and that the greatest philosophers of the present day know less of it than of any other physical phenomenon.

APPENDIX.

APPENDIX.

LIST OF THE PRINCIPAL WORKS THAT HAVE
CONTRIBUTED TO OUR PRESENT KNOWLEDGE
OF PHOSPHORESCENCE.

FABRICIO (Jérôme) *d'Aquapendente.* On the Phosphorescence
of Flesh, in *Opera Omnia.* Edition published at Leipsic in
1687.

BARTHOLIN (Thomas). *De Luce Hominum et Brutorum.* Published at Leyden in 1647. But two other editions were published in 1663 and 1669, in 8vo (Haffniæ) ; and to the latter
has been added the treatise of Gesner, entitled *De Raris et
Admirandis Herbis quæ Noctu Lucent.*

LICETUS. *Fortunii Liceti Litheosphorus, sive de Lapide Bono-
niensi.* (Utini, 1640.) Account of the Discovery of the Bologna Stone.

POTIER (Pierre). *Pharmacopœia Spagyrica,* ii. 27, in *Opera*
(Frankfort, 1698.) Description of the method formerly em-
ployed for preparing Solar Phosphorus from the Bologna
Stone.

MARGGRAF. *Chymische Scriften,* vol. ii. New Method for preparing Solar Phosphorus.

BOYLE (Robert). *On the Phosphorescence of the Diamond,
Adamas Lucens, &c.* (Lond. 1663.) "Experiments concerning
the relation of air and light in shining wood and fish :" *Philosophical Transactions,* 1668. "Observations on shining
flesh :" *Phil. Trans.,* 1672. "Account of the new element
Phosphorus :" *Phil. Trans.,* 1680.

LEMERY (Nicholas). Prep. of Solar Phosph.: *Cours de Chymie.* (1675 ; the best edition of this remarkable work was published at Paris in 1713.) Lemery's knowledge of Phosphorescence was greater than that of any man of his time. The pages 482, 483 *et seq.* of his work on Chemistry, may be read with interest at the present day. It is exceedingly remarkable that his explanation, constantly repeated, of the *light being owed to rapid molecular motion,* is precisely that which is becoming adopted at present.

BAUDUIN. (Some authors write his name Balduin, and Baldwin.) On his Phosphorus produced from Nitrate of Lime. *Phosphorus Hermeticus, seu magnes Luminaris,* 1675; see also *Phil. Trans. Abrid.* ii. p. 368.

HOMBERG. On his Phosphorus. *Mémoires de l'Académie de Paris,* 1693, p. 307. He prepared it by heating sal-ammoniac and chalk. The result was carbonate of ammonia which distilled and dry-fused *chloride of Calcium,* which shines in the dark when struck.

HAUKSBEE. On the Luminosity produced by the Friction of Mercury in the Barometer Vacuum. *Physico-mechanical Experiments.* (London, 1709.) "Experiments on the attrition of bodies *in vacuo : Phil. Trans.,* 1705. In this paper the author proves by direct experiment, that when steel is rubbed against flint, the sparks are not produced without the presence of air ; hence it was afterwards discovered that, in this case, the light is produced by the rapid combustion of small particles of steel.

BOURYES (Father). "Concerning the Luminous Appearance observable in the Wake of Ships," &c.: *Phil. Trans.,* 1713. I have not alluded to this paper in my historical notes, as the author brings forward no observations of particular value, and has not the most remote idea of the true cause of this phosphoric appearance. "The production of the light," he says, "depends very much on the quality of the water." The author's observations are however interesting, as having been made in the Indian Seas.

Dufay. On the Phosphorescence of Diamonds, etc.: *Mémoires de l'Acad. de Paris*, 1730, and *id.* 1735. According to this author, yellow diamonds are more apt to become phosphorescent than those of other tints.

Wilson. Influence of the different rays of the Solar Spectrum on Phosphoric Light: *Journ. de Physique*, t. xv. p. 92.

Beccaria. *De quamplurimis Phosphoris, etc.* (Bologna, 1744), and on luminous clouds, etc., quoted by Arago in the *Annuaire* for the year 1838.

Vianelli. *Nuove Scoperte intorno le Luci notturne dell' Acqua Marina.* (Venezia, 1749.)

Grixellini. On the Luminous Scolopendra. (Venice, 1750.) I am not acquainted with this work.

Canton. On his Phosphorus, in *Phil. Trans.*, 1768.

Wedgwood (Thomas). " Experiments and Observations on the Production of Light from different bodies by Heat and by Attrition :" *Phil. Trans.*, 1792.

Hulme. Exp. on the Light which is spontaneously emitted from various bodies, and on Solar Light: *Phil. Trans.*, 1802. In this remarkable paper Dr. Hulme states, " that solar light, when imbibed by Canton's Phosphorus, is subject to the same laws with respect to heat and cold as the spontaneous light of fishes, rotten wood, and glowworms."

Viviani. *Phosphorentia Maris*, etc. (Genoa, 1805.)

Dessaignes. Papers on Phosphorescence, in *Mémoires de l'Académie de Paris*, 1807, 1808.

Pearsal. " On Phosphorescence by Heat," etc.: *Ann. de Chim.* 2nd series, t. xlix. pp. 337, 346. " Electricity and Phosphorescence :" *Journ. of the Royal Institution*, vol. i.

Brewster. " On Phosphorescence by Heat :" *Edin. Phil. Mag.* i. 383.

Heinrich (Placidus). " On Phosphorescence after Insolation and by Heat :" *Journ. de Phys.* lxxi. 307.

Seebeck. " Exp. on Phosphorescence :" *Comptes-Rendus de l'Acad. des Sc. de Paris*, t. xiv.

Grothus. " Exp. on Phosphorescence :" *Schweiger's Journal*, xiv. p. 134.

MACARTNEY. "On Luminous Animals:" *Phil. Trans.*, 1810.

THOMSON (Thos.) *A System of Chemistry*, in four vols. Vol. i. chap. 1, of Light. (London, 1820.)

SOMERVILLE (Mary). "On the Magnetizing Power of the more refrangible Solar Rays:" *Phil. Trans.*, 1826. MORI-CHINI's Experiments were published in *Gilbert's Annalen der Physik*, in 1813; see also *Annals of Philosophy*, ii. 390, where PLAYFAIR gives an account of the success of these experiments in the hands of M. CAUPE. In *Ann. of Phil.*, iv. 228, they are, however, denied by other authors.

MICHAELIS. *Ueber das Leuchten der Ostsee.* (Hamburg, 1830.)

TREVIRANUS. *Die Erscheinungen und Gesetze des Thierischen Lebens.* (Bremen, 1831.) And his *Vermischte Schriften*, edited with his brother from 1816 to 1821. 4 vols. (Bremen and Göttingen.)

TILESIUS. On the Phosphorescence of Small Medusæ, in *Annalen der Wetterauischen Gesellschaft*, vol. iii. p. 360. A paper quoted by CARUS in his *Comp. Anat.*, vol. i. (Belg. ed. 1838.)

PONTUS. On the Spark produced by Water when it is made to freeze: *Journ. des Sciences Physiques de M. Julia de Fontanelle*, vol. i. p. 131. (1833.)

DELILLE. On the Phosphorescence of the Italian Agaric, in *Comptes-Rendus of the Acad. des Sc. Paris*, 1833.

DUMAS. Discovery of the Phosphoric Radiation emitted by Boracic Acid when cooling after melting (cited by BERZELIUS). *Traité de Chimie* (Belg. ed.), vol. i.

BERZELIUS. Discovery of the light produced when fluoride of sodium crystallizes, and other cases of mineral phosphorescence, in *Traité de Chimie*, vol. i. to v. (Belg. ed.)

HARRIS (Sir W. Snow). On the Investigation of Magnetic Intensity, etc., in *Edin. Phil. Trans.*, January, 1834. In this paper Sir William Snow Harris has arrived at the conclusion that magnetism induced by the sun's rays does not occur in a vacuum. "The influence of the sun's rays," says the author, "on a magnet oscillating in air, is to reduce more rapidly the

arc, and to diminish the time of a given number of vibrations.
The influence of the sun's rays on a magnet oscillating in a
void, is to increase the time of a given number of vibrations;
whilst the arc, if the retardation of the rate of vibration be
small, is not materially affected." Nearly the same phenomena occur when bars of copper are vibrated in like manner.

MELLONI. "Observations et Expériences relatives à la théorie
et à l'identité des agents qui produisent la lumière et la cha-
leur rayonnante," in *Journ. des Sc. Phys. de M. Julia de Fontanelle*, t. iv. p. 161. (1836.) In this remarkable paper, which
was, I think, presented to the Academy of Sciences of Paris,
the author, after proving that radiant light and heat have two
distinct causes, or are two distinct effects of one cause, endeavours to separate light completely from heat. His experiments
succeeded admirably. Absorbing all the heat either of the
solar rays or an artificial fire, by making the rays pass through
a system of diaphanous bodies, such as water or blue glass, he
succeeded in obtaining *pure light*. "This *pure light*," says
Melloni, "emerging from this system, contains much *yellow*,
and possesses moreover a *bluish-green tint*. It has no action
whatever on the most delicate thermometer, even when condensed by a lens until it is as brilliant as the direct rays of the
sun." This *pure light* has therefore all the properties hitherto
recognized in ordinary phosphoric light, such as that of the
glowworm, the Bologna-stone, etc.

SURIRAY. "Recherches sur la cause ordinaire de la Phosphorescence Marine."—*Mag. de Zoologie de Guérin*, 1836.

BECQUEREL (Ed.). On Phosphorescence after Insolation; many
papers in *Ann. de Chim. Paris.* Compare also MORREN, in
Comptes-Rendus, 1861 and 1862, on the Phosphorescence of
Gases.

BECQUEREL (Père). *Traité d'Electricité*, t. vi.

MOIGNO. *Repert. d'Optique Moderne.* (Paris, 1842.)
These two authors have quoted several facts relating to Phosphorescence in their large works.

ARAGO. "Notice sur le Tonnerre," etc., in his *Œuvres Complètes.*

Marchand. "On the Phosphorescence of Phosphorus :" *Journ. für Prakt. Chemie,* 1851, l. p. 1. We have here a set of experiments, tending to show that the phosphorescence of phosphorus is not owing to combustion, but to a molecular action upon the surface of this substance, and which manifests itself in all kinds of gases, whether they support combustion or not. But the phosphorescence does not appear to be of long duration in gases, such as carbonic acid and hydrogen, and moreover depends upon the pressure to which the gases are submitted. Marchand thinks that phosphorus is capable of shining in the dark in carbonic acid, hydrogen, etc., and that the phenomenon is owing to the volatilization going on at the surface of the phosphorus.

Schroetter. In 1852, Professor Schroetter of Vienna, after assuring himself by several experiments that the light given out by *Phosphorus,* when its temperature is slightly raised, is owing to oxidation, showed that *Sulphur, Selenium, Tellurium,* and *Arsenic,* heated gradually in contact with oxidizing bodies, give out light and produce oxides which differ from those produced by ordinary combustion at higher temperatures.—*Le Cosmos,* vol. i. (Paris, 1852.)

Ehrenberg. *Ueber das Leuchten des Meeres. Abhand. der Akad. zu Berlin,* 1854.

De Quatrefages. On the Phosphorescence of Noctiluca : *Comptes-Rendus,* t. xxxi., and *Annales des Sciences Naturelles,* 3ᵉ série, vol. xiv.

Verhaeghe. On the Phosphorescence of the Sea at Ostend. A pamphlet reprinted from the *Bulletin de l'Acad. des Sc. de Bruxelles,* 1855.

Rose (Heinrich). On the emission of light by Arsenious Acid whilst crystallizing : *Ann. der Physik,* 1835, and *Ann. de Chim.,* 2ᵉ série, lxi. 288. Light developed during the crystallization of sulphate of soda and potash : *Ann. de Chim.,* 3ᵉ série, lv. 125. On the phosphorescence of certain substances when heated : *Annales der Physik,* March, 1858, and *Ann. de Chim.,* January, 1859.

PHIPSON. "Sur une espèce particulière de mucilage animale:" *Journ. de Médecine et de Pharm. de Bruxelles,* 1855.—*De la Phosphorescence en général et des Insectes phosphoriques.* Bruxelles and Paris, 1858. Abridg. ed. in German, published at Berlin, 1858, by Dr. Müller.—On some new cases of Phosphorescence by Heat, in *Comptes-Rendus de l'Acad. des Sc. de Paris,* February, 1860.—*Recherches nouvelles sur le Phosphore.* Brussels, 1856.—*La Force Catalytique, Etudes sur les Phénomènes de Contact. Mémoire couronné par la Société Hollandaise des Sciences, Harlem.* 1858.— *Sur la Matière Phosphorescente de la Raie: Comptes-Rendus,* 1860.

NIÉPCE DE ST. VICTOR. On a new Action of Light. Many papers in *Comptes-Rendus de l'Acad. des Sc. Paris,* from 1857 to 1858.

BECQUEREL (Edmond). On Phosphorescence after Insolation: *Ann. de Chim.,* January, 1859, and some papers on the same subject in *Comptes-Rendus* since the date just given.—On the Phosphorescence of Gases: *Comptes-Rendus de l'Acad. des Sc. Paris,* 1859.

WARTMANN. On a Luminous Fog at Geneva: *Comptes-Rendus,* December, 1859.

The names of other authors have been given in the present work.

Note.—It has been observed that *Calamine* (hydrated silicate of zinc), when heated, becomes *electric* and *phosphorescent* at the same time; and this will doubtless be found to be the case with many other mineral substances which exhibit phosphorescence.

The mineral called *Wollastonite* (silicate of lime) becomes phosphorescent by friction.

My attention has lately been called to the fact that *ice* and *snow* are phosphorescent after insolation. Although I have not had an opportunity of witnessing it myself, I find that Placidus Heinrich was acquainted with the phenomenon; he

P

found that large lumps of ice are slightly phosphorescent in a dark room after being exposed to the sun, the temperature being kept several degrees below freezing point. The brothers Schlagintweit, Prof. Berty, and Mr. Tuckett, have witnessed the phosphorescence of the snow and the ice on the glaciers of the Alps, etc. The darker the night, the more brilliant the phenomenon. It appears sometimes like the effect of a second sunset. This phosphorescence is remarked on the Alpine summits and on the snow which lies in the valleys of Piedmont, Switzerland, Valais, etc. The colour of the light emitted is bluish. It is not remarked in snow which has fallen shortly before night, and which, consequently, has not been long exposed to the sun.

THE END.

JOHN EDWARD TAYLOR, PRINTER,
LITTLE QUEEN STREET, LINCOLN'S INN FIELDS.

www.ingramcontent.com/pod-product-compliance
Lightning Source LLC
Chambersburg PA
CBHW021556210326
41599CB00010B/465